Quantitative Measure for
Discrete Event Supervisory Control

T0191988

Quantitative Measure for Discrete Event Supervisory Control

edited by

Asok Ray
The Pennsylvania State University
University Park, PA, USA

Vir V. Phoha
Louisiana Tech University
Ruston, LA, U.S.A.

Shashi P. Phoha
N.I.S.T. –Information Technology Laboratory
Gaithersburg, MD, USA

Asok Ray
The Pennsylvania State Univ.
University Park, PA

Vir V. Phoha
Louisiana Tech Univ.
Ruston, LA

N.I.S.T.
Information Technology Lab.
Gaithersburg, MD 20899

Shashi P. Phoha

Library of Congress Cataloging-in-Publication Data

A C.I.P. Catalogue record for this book is available from the Library of Congress.

Quantitative Measure for Discrete Event Supervisory Control / Edited by Asok Ray, Vir V. Phoha, Shashi P. Phoha
    p.cm
    Includes bibliographical references and index.

ISBN 978-1-4419-1851-2 Printed on acid-free paper.

1. Supervisory control systems. 2. Discrete-time systems. I. Ray, Asok, 1946- II. Phoha, Vir V. III. Phoha, Shashi, 1948-

TJ222.Q36 2005
629.8—dc22

2004059173

© 2005 Springer Science+Business Media, Inc.
Softcover reprint of the hardcover 1st edition 2005

9 8 7 6 5 4 3 2 1
11355632 (eBook)

SPIN 10925904 (HC) /

springeronline.com

Contents

Asok Ray dedicates this monograph to his students, teachers,
and colleagues
Vir Phoha dedicates this monograph to his parents and his
brother Rajesh
Shashi Phoha dedicates this monograph to her parents

Preface

Discrete event systems belong to a special class of dynamical systems. The states of a discrete event system may take discrete (or symbolic) values and change only at (possibly asynchronous) discrete instants of time, in contrast to the familiar continuously varying dynamical systems of the physical world, which can be modelled by differential or difference equations. The dynamics of many human-engineered systems evolve asynchronously in time via complex interactions of various discrete-valued events with continuously varying physical processes. The relatively young discipline of discrete event systems has undergone rapid growth over the last three decades with the evolution of human engineered complex systems, such as integrated control and communication systems, distributed sensing and monitoring of large-scale engineering systems, manufacturing and production systems, software fault management, and military Command, Control, Computer, Communication, Intelligence, Surveillance, and Reconnaissance (C^4ISR) systems.

The discipline of discrete event systems was initiated with simulation of human-engineered processes about four decades ago in the middle of nineteen sixties. The art of discrete event simulation emerged with the development of a simulation software package, called GPSS, that was followed by numerous other software simulation tools, such as SIMSCRIPT II.5, SLAM II, and SIMAN [10]. Shortly thereafter, computer scientists and control theorists entered the field and brought in theoretical concepts of languages and automata in modelling discrete event systems. In the late nineteen sixties, Arbib [8] showed how algebraic methods could be used to explore the structure of finite automata to model dynamical systems. Around that time, computer scientists focused on formal languages, automata theory, and computational complexity for application of language-theoretic concepts (e.g., regular expressions and context-free grammars) in software development including design of compilers and text processors [7]. In the late nineteen seventies and early nineteen eighties, Ho and co-workers introduced the concept of finite perturbation in discrete event systems for modelling and analysis of human-engineered systems [5]. So far, no concrete theoretical concept and mathematical tools had been available for analysis and synthesis of discrete event control systems.

The concept of discrete event supervisory (DES) control was first introduced in the seminal paper of Ramadge and Wonham [12] and this important paradigm has been subsequently extended by other researchers (for example, see citations in [9] [1], and the October 2000 issue of Part B of IEEE Transactions on Systems, Man, and Cybernetics). These efforts have led to the evolution of a new discipline in decision and Control, called Supervisory Control Theory (SCT), that requires partitioning the discrete-event behavior of a physical process, called the plant, into legal and illegal categories. The legal behavior of the plant dynamics is modelled by a deterministic finite-state automaton, abbreviated as DFSA in the sequel. The DFSA model is equivalent to a regular language that is built upon an alphabet of finitely many events; the event alphabet is partitioned into subsets of controllable events (that can be disabled) and uncontrollable events (that cannot be disabled). Based on the regular language of an unsupervised plant, SCT synthesizes a DES controller as another regular language, having the common alphabet with the plant language, that guarantees restricted legal behavior of the controlled plant based on the desired specifications. Instead of continuously handling numerical data, DES controllers are designed to process event strings to disable certain controllable events in the physical plant. A number of algorithms for DES control synthesis have evolved based on the automata theory and formal languages relying on the disciplines of Computer Science and Control Science. In general, a supervised plant DFSA is synthesized as a parallel composition of the unsupervised plant DFSA and a supervisor DFSA. The supervised plant DFSA yields a sublanguage of the unsupervised plant language, which enables restricted legal behavior of the supervised plant [12] [9] [1]. These concepts have been extended to several practical applications, including hierarchical Command, Control, Communications, and Intelligence (C^3I) systems [11]. Apparently, there have been no quantitative methods for evaluating the performance of supervisory controllers and establishing thresholds for their performance.

The concept of permissiveness has been used in DES control literature [1] [9] to facilitate qualitative comparison of DES controllers under the language controllability condition. Design of maximally permissive DES controllers has been proposed by many researchers based on different conditions. However, maximal permissiveness does not necessarily imply best performance of the supervised plant from the perspectives of plant operational objectives. For example, in the travelling salesman problem, a maximally permissive supervisor would not usually yield the least expensive way of visiting the scheduled cities and returning to the starting point because no quantitative measure of performance is addressed in this type of supervisor design. Hence, permissiveness is not an adequate measure of performance.

The controlled behavior of a given plant DFSA under different supervisors could vary if they are designed to meet different control specifications. As such the respective controlled sublanguages of the plant language form a partially ordered set that is not necessarily totally ordered. Since the literature on DES control does not apparently provide a language measure, it may not be possible to quantitatively evaluate the performance of a DES controller. Therefore, it is necessary to formulate a mathematically rigorous concept of language measure(s) to quantify performance of individual supervisors such that the measures of controlled plant behavior, described by a partially ordered set of controlled sublanguages, can be structured to form a totally ordered set. From this perspective, it is necessary to formulate a signed real measure μ that can be assigned to sublanguages of the regular language of the unsupervised plant to achieve the following objective [16] [15] [13]:

> *Given that the relation $\subseteq$ induces a partial ordering on a set of controlled sublanguages $\{L(S^j/G), j = 1, \ldots, N\}$ of the plant language $L(G)$ under supervisors whose languages are $\{L(S^j), j = 1, \ldots, N\}$, the language measure μ induces a total ordering $\leq$ on $\{\mu(L(S^j/G))\}$.*

The major motivation here is to present a signed real measure of regular languages, which can be used for quantitative analysis and synthesis of DES control systems for physical plants, instead of relying on permissiveness as the (qualitative) performance index. Construction of the proposed language measure follows Myhill-Nerode Theorem [6] and Hahn Decomposition Theorem [14], where a state-based partitioning of the (unsupervised) plant language yields equivalence classes of finite-length event strings. Each marked state is characterized by a signed real value that is chosen based on the designer's perception of the state's impact on the system performance. Conceptually similar to the conditional probability, each event is assigned a cost based on the state at which it is generated. This procedure permits a string of events, terminating on a good (bad) marked state, to have a positive (negative) measure; each of the remaining strings of events in the language, terminating on an unmarked state, is assigned zero measure. In this setting, a supervisor can be designed with the goal of eliminating bad strings and retaining good strings by disabling controllable event(s) at selected states. Different supervisors may attempt to achieve this goal in different ways and generate a partially ordered set of controlled languages. The language measure then creates a total ordering on the performance of the controlled languages, which provides a precise quantitative comparison of the controlled plant behavior under different supervisors.

The language measure has been used as a performance index for analysis and synthesis of DES control systems. For example, Fu et al. [4] [3] [2] have proposed optimal control and robust optimal control of regular languages where a state-based optimal control policy is obtained by selectively disabling controllable events to maximize the measure of the controlled plant language.

The concept of DES control has been validated by laboratory experimentation on mobile robots as well as on simulation test beds for applications to gas turbine engines, software fault management, C^3I systems, and control of malicious executables in computers,

Main Features of the Research Monograph

This monograph presents a novel and mathematically rigorous approach to discrete event supervisory control, and builds on the research work, some of which is only recently available in conference proceedings, or will be available in journals within one to two years because of lag time in publication. The new concepts and their applications are presented for practicing professionals and graduate students in engineering, computer science, physics, applied mathematics as well as in other fields such as economics, life sciences, social sciences, and political sciences. It is envisioned that the presented inter-disciplinary material will significantly enhance the science and technology base not only in engineering and physical sciences but also in other disciplines such as management sciences, biological sciences, social sciences, and economics.

Starting from the fundamental principles of automata theory and languages, this research monograph presents a *quantitative measure of formal languages*, which lays the theoretical foundations for supervisory decision and control in complex human-engineered systems with application examples of anomaly detection and fault handling in robotics, gas turbine engines, and software systems. It introduces a new paradigm to quantitatively compare the performance of different automata models of a controlled physical process. The language measure is well suited to serve as a performance index for optimal and robust optimal supervisory decision algorithms in the discrete-event setting for both military and commercial C^3I systems. This approach, which was first reported in the Proceedings of 2002 American control Conference and is not available in any existing monographs and textbooks, will help disseminate the underlying concepts and spread its use in various disciplines and applications.

To the best of the editors' knowledge, this monograph is the first comprehensive, self-contained treatment of a language-theoretic quantitative approach to discrete event supervisory decision and control. It complements the information contained in other research monographs (e.g., [9]) and text books (e.g., [1]) on discrete event supervisory control. The main features of this monograph are summarized below:

- The monograph focuses on quantification of regular languages as a signed real measure. It provides insight and understanding through rigorous mathematical foundations and necessary proofs, while avoiding non-essential mathematical details and emphasizing applications. The concept of signed real measure of formal languages, which is introduced in this monograph, either has been very recently reported by the authors in research publications and conference proceedings, or are yet to be published in open literature. The language measure provides a formal way to quantitatively compare different supervisors for a given plant.

- The monograph presents a quantitative approach to analysis and synthesis of optimal and robust supervisory decision and control systems in the discrete-event setting.
- The monograph identifies advanced research topics in supervisory control beyond what is practiced in the current state of the art.
- The monograph presents novel applications of the language measure theory: Real-time supervisory decision and control for fault management in engineering systems.
- Parts of the monograph have been classroom tested in Pennsylvania State University and Louisiana Tech University; its comprehensive bibliography serves as a rich research resource.

Audience

This monograph is intended primarily for researchers and graduate students in engineering and science. It should also be useful to practicing professionals. The reader is assumed to have (undergraduate level) mathematical background in calculus, matrix theory, and abstract algebra. The applications-oriented parts of the monograph require domain-specific knowledge. The monograph presents the concepts and their applications to be easily comprehensible to students in engineering, computer science, mathematics, and physics. Students in other disciplines, such as biology, political science, economics, sociology, and anthropology, may require limited pre-requisites on undergraduate mathematics. It is envisioned that the inter-disciplinary materials, presented in this monograph, will significantly expand the knowledge base of students not only in Engineering and Science but also in other disciplines.

Organization

The technical content of the monograph is organized into two parts: Part I – Theory of Language Measure and Supervisory Control; and Part II – Engineering and Software Applications of Language Measure and Supervisory Control. Part I consists of four chapters and Part II consists of five chapters. Each chapter is self-contained as it briefly reviews the prerequisite materials at the beginning. This will allow the readers to go through any individual chapter without the burden of first reading previous chapter(s) and other citations; however, the readers who are interested in more details are encouraged to refer to the appropriate previous chapter(s) and cited references.

Part I: Theory of Language Measure and Supervisory Control

The first part has four chapters focusing on the theory of language measure, and optimal and robust supervisory control in the discrete event setting. These four chapters report current research work and future research topics that have recently appeared in the proceedings of national and international conferences or will appear in archival journals.

Chapter 1 formulates a signed real measure of sublanguages of a regular language based on the principles of automata theory and real analysis. The measure allows total ordering of any set of partially ordered sublanguages of the regular language for quantitative evaluation of the controlled behavior of deterministic finite state automata (DFSA) under different supervisors. In the setting of the language measure, a supervisor's performance is superior if the supervised plant is more likely to terminate at a *good* marked state and/or less likely to terminate at a *bad* marked state. The computational complexity of the language measure algorithm is polynomial in the number of DFSA states.

Chapter 2 presents optimal supervisory control of dynamical systems that can be represented by deterministic finite state automaton (DFSA) models. The performance index for the optimal policy is obtained by combining a measure of the supervised plant language with (possible) penalty on disabling of controllable events. The signed real measure quantifies the behavior of controlled sublanguages based on a state transition cost matrix and a characteristic vector as reported in Chapter 1 and recent literature. Synthesis of the optimal control policy requires at most n iterations, where n is the number of states of the DFSA model generated from the unsupervised plant language. The computational complexity of optimal control synthesis is polynomial in n. Syntheses of the control algorithms are illustrated with two application examples.

Chapter 3 presents an algorithm for robust optimal control of regular languages under specified uncertainty bounds for the event costs of a language measure. This algorithm has been recently reported in literature and is presented in Chapter 1. The performance index for the proposed robust optimal policy is obtained by combining the measure of the supervised plant language with uncertainty. The performance of a controller is represented by the language measure of the supervised plant and is minimized over the given range of event cost uncertainties. Synthesis of the robust optimal control policy requires at most n iterations, where n is the number of states of the deterministic finite state automaton (DFSA) model generated from the regular language of the unsupervised plant behavior. The computational complexity of control synthesis is polynomial in n.

Chapter 4 introduces two research topics in the field of language-measure-based supervisory control. One topic is complex measure of non-regular languages starting with *linear context free grammars* (*LCFG*), where the proposed complex measure is reduced to the signed real measure if the *LCFG* is degenerated to a regular grammar. The other topic is a modification of the (regular) language measure for supervisory control under partial observation. While the theory of optimal supervisory control of processes is presented in earlier chapters for completely observable finite state automata, this chapter shows how to generalize this concept to situations where some of the events may not be observable at the supervisory level.

Part II: Engineering and Software Applications of Language Measure and Supervisory Control

The second part that demonstrates applications of the theoretical work, reported in Part I of the monograph, consists of five chapters. Several case studies in physical processes (e.g., a group of mobile robots, and gas turbine engines) and human-engineered complex systems (e.g., computer operating systems, and malicious executables) are presented in these chapters.

Chapter 5 presents an application of the theory of Discrete Event Supervisory (DES) control, presented in Chapters 1 and 2, and addresses the design of a robotic system interacting with a dynamically changing environment. The work, reported in this chapter, encompasses the disciplines of control theory, signal analysis, computer vision, and artificial intelligence. Several traditional important methods are first reviewed to substantiate the DES control approach in the design of a mobile robotic system. Design and modelling of the behavior-based mobile robotic system is then presented in details. The plant automaton model of the robotic system is identified by making use of the available sensors and actuators. Then, a DES controller is synthesized based on the data collected from experimental scenarios. Through these experiments, performance of the robotic DES control system is quantitatively evaluated in terms of the language measure for both the unsupervised and supervised systems. It is shown that the language measure can indeed be used as a performance index in the design of optimal DES control policies for higher level mission planning.

Chapter 6 presents discrete-event supervisory control of robot behavior in terms of the language measure, presented in Chapter 1. In the discrete-event setting, a robot's behavior is modelled as a regular language that can be realized by deterministic finite state automata (DFSA). Pertinent parameters of the robot behavior language are identified either from experimental data or from the results of extensive simulation, as they are dependent on physical phenomena. Since the robot behavior is often slowly time-varying, there is a need for on-line parameter identification to generate up-to-date values of the language parameters within allowable bounds of errors.

Chapter 7 presents an application of the theory of optimal Discrete Event Supervisory (DES) control that is based on a signed real measure of regular languages described in Chapter 1. The DES control techniques are validated on an aircraft gas turbine engine simulation test bed. The test bed is implemented on a networked computer system in which two computers operate in the client-server mode. Several DES controllers have been tested for engine performance and reliability. Extensive simulation studies on the test bed show that the optimally designed supervisor yields the best performance.

Chapter 8 presents a novel approach to anomaly and fault management of software systems. The traditional approach to fault tolerance identifies anomalies and modifies the software application to mitigate the detrimental effects of the anomalies; the core algorithm, presented in this chapter, is built upon Supervisory Control Theory (SCT) to monitor the execution of software applications by restricting the actions of the underlying Operating System (OS) to mitigate the effects of observed anomalies. The proposed approach can be generalized to any software application; its interactions with the OS are modelled at the process level as a Deterministic Finite State Automaton (DFSA), called the *Plant*, which is synchronously controlled by another DFSA, called the *Supervisor*. The Supervisor is synthesized to restrict the language accepted by the plant to satisfy the control specifications for anomaly and fault mitigation. As a proof-of-concept, two supervisors are implemented under the Redhat Linux 7.2 OS to mitigate overflow and segmentation faults in five different programs. The performance of the unsupervised and supervised plant is quantified by making use of the language measure, presented in Chapter 1; the language measure parameters are evaluated from experimental data.

Chapter 9 models the execution of a software process as a discrete event system that can be represented by a Deterministic Finite State Automaton (DFSA) in the discrete event setting. Supervisory Control Theory (SCT) is applied on-line to detect and prevent spreading of malicious executables. Then, the language measure theory, described in Chapter 1, is adapted to evaluate performance of the unsupervised and supervised processes. Simulation studies under different scenarios are presented to show how a supervisor detects malicious executables and controls their spreading on-line. The simulation results show the rate of correct detection to be 88.75% and thus demonstrate the potential of discrete-event supervisory (DES) control to prevent spreading of malicious executables.

Acknowledgements

This work has been supported in part by Army Research Office (ARO) under Grant No. DAAD19-01-1-0646; Defense Advanced Research Projects Agency (DARPA) and Air Force Research Laboratory, Air Force Materiel Command, USAF, under Agreement No. F3062-01-0575; NASA Glenn Research Center under Grant Nos. NAG3-2448 and NNC04GA49G; and National Science Foundation under Grant Nos. CMS-9819074, ECS-9912495 and ECS-9984260.

The editors and authors acknowledge contributions of their students and colleagues for enhancement of the quality of the reported work.

State College, Pennsylvania   *Asok Ray*
Ruston, Louisiana   *Vir V. Phoha*
State College, Pennsylvania   *Shashi Phoha*

References

1. C.G. Cassandras and S. Lafortune, *Introducrion to discrete event systems*, Kluwer Academic, 1999.
2. J. Fu, C.M. Lagoa, and A. Ray, *Robust optimal control of regular languages with event cost uncertainties*, Proceedings of IEEE Conference on Decision and Control, December 2003, pp. 3209–3214.
3. J. Fu, A. Ray, and C.M. Lagoa, *Optimal control of regular languages with event disabling cost*, Proceedings of American Control Conference, Denver, Colorado, June 2003, pp. 1691–1695.
4. J. Fu, A. Ray, and C.M. Lagoa, *Unconstrained optimal control of regular languages*, Automatica **40** (2004), no. 4, 639–648.
5. Y-C. Ho and X-R. Cao, *Perturbation analysis of discrete event dynamic systems*, Kluwer Academic, 1991.
6. J. E. Hopcroft, R. Motwani, and J. D. Ullman, *Introduction to automata theory, languages, and computation, 2nd ed.*, Addison-Wesley, 2001.
7. J. E. Hopcroft and J. D. Ullman, *Introduction to automata theory, languages, and computation, 1st ed.*, Addison-Wesley, 1979.
8. R.E. Kalman, P.L. Falb, and M.A. Arbib, *Topics in mathematical system theory*, McGraw-Hill, 1969.
9. R. Kumar and V. Garg, *Modeling and control of logical discrete event systems*, Kluwer Academic, 1995.
10. A.M. Law and W.D. Kelton, *Simulation modeling & analysis, 2nd ed.*, McGraw-Hill International, 1991.
11. S. Phoha, E. Peluso, and R.L. Culver, *A high fidelity ocean sampling mobile network (samon) simulator*, IEEE Journal of Oceanic Engineering, Special Issue on Autonomous Ocean Sampling Networks **26** (2002), no. 4, 646–653.
12. P.J. Ramadge and W.M. Wonham, *Supervisory control of a class of discrete event processes*, SIAM J. Control and Optimization **25** (1987), no. 1, 206–230.
13. A. Ray and S. Phoha, *Signed real measure of regular languages for discrete-event automata*, Int. J. Control **76** (2003), no. 18, 1800–1808.
14. W. Rudin, *Real and complex analysis, 3rd ed.*, McGraw-Hill, New York, 1987.
15. A. Surana and A. Ray, *Signed real measure of regular languages*, Demonstratio Mathematica **37** (2004), no. 2, 485–503.
16. X. Wang and A. Ray, *A language measure for performance evaluation of discrete-event supervisory control systems*, Applied Mathematical Modelling **28** (2004), no. 9, 817–833.

About the Editors

There are three editors of this book, Drs. Asok Ray, Vir V. Phoha, and Shashi Phoha. Brief biographical sketches of the editors follow.

Dr. Asok Ray

Dr. Asok Ray, Distinguished Professor of Mechanical Engineering, Pennsylvania State University. Previously worked at Carnegie Mellon University, Massachusetts Institute of Technology, Charles Stark Draper Laboratory, and MITRE. Fellow of IEEE; Fellow of ASME; Associate Fellow of AIAA; Professional Engineer. Editor of IEEE Transactions on Aerospace and Electronic Systems; Editorial Board Member, Advances in Industrial Control Series, Springer-Verlag, London; Associate editor of IEEE Transactions on Control Systems Technology; Associate Editor, The International Journal of Structural Health Monitoring (IJSHM); Associate Editor, International Journal of Flexible Manufacturing Systems; Past Associate Editor, ASME Journal of Dynamic Systems, Measurement, and Control (1992-1995). Author of over 400 research publications and book chapters including a Springer-Verlag, London monograph, Intelligent Seam Tracking for Robotic Welding (1993). More details available at http://www.me.psu.edu/Ray/.

Dr. Vir V. Phoha

Dr. Vir V. Phoha, Associate professor of Computer Science in the College of Engineering and Science at Louisiana Tech University. Awarded various distinctions including outstanding research faculty and faculty circle of excellence at Northeastern State University, Oklahoma, and, as a student, President's medal for academic distinction. Author of over 45 research publications and two books: (1) Internet Security Dictionary, Springer-Verlag (2002) and (2) Foundations of Wavelet Networks and Applications, CRC Press/Chapman Hall (2002).

Dr. Shashi Phoha

Dr. Shashi Phoha, Director of Information Technology Laboratory at National Institute of Standards and Technology (NIST), and Professor of Electrical and Computer Engineering at Pennsylvania State University. Held senior technical and management positions at Computer Sciences Corporation, ITT Defense Communications Division, and MITRE Corporation. Established major advanced research projects in industry and academia. Founder and Director of a University/Industry consortium for establishing a National Information Infrastructure Interoperability Testbed, sponsored by Defense Advanced Projects Agency (DARPA) and Principal Investigator of several DoD MURI Programs. Editor of International Journal of Distributed Sensor Networks and Associate Editor of IEEE Transactions on Systems, Man, and Cybernetics. Member of the Board of Directors of International Consortium CERES Global Knowledge Network, along with representatives of thirteen other international universities. Served on the Board of Directors of Autonomous Undersea Vehicle Technology Consortium for International Cooperation Between Research, Technology, Industry and Applications. Member of the National Information Infrastructure Standards Panel (ANSI). Author of over 150 scholarly articles and book chapters, and two books.

Part I

Theory of Language Measure and Supervisory Control

1

Signed Real Measure of Regular Languages

Asok Ray[1] and Xi Wang[2]

[1] The Pennsylvania State University axr2@psu.edu
[2] The Pennsylvania State University xxw117@psu.edu

Summary. This chapter formulates a signed real measure of sublanguages of a regular language based on the principles of automata theory and real analysis. The measure allows total ordering of a set of partially ordered sublanguages of the regular language for quantitative evaluation of the controlled behavior of deterministic finite state automata (DFSA) under different supervisors. In the setting of the language measure, a supervisor's performance is superior if the supervised plant is more likely to terminate at a *good* marked state and/or less likely to terminate at a *bad* marked state. The computational complexity of the language measure algorithm is polynomial in the number of DFSA states.

Key words: Discrete Event Supervisory Control, Performance Measure, Formal Languages

1.1 Introduction

Discrete-event dynamic behavior of a physical plant is often modelled as a deterministic finite-state automaton (DFSA) that can be represented by a regular language [7] [9]. The concept of Discrete Event Supervisory (DES) control, pioneered by Ramadge and Wonham [14] and subsequently enhanced by other researchers (see, for example, citations in [1]), provides a framework for achieving prescribed qualitative performance of physical plants. A parallel composition of the unsupervised plant automaton and a supervisor automaton yields a sublanguage of the unsupervised plant language, which enables restricted legal behavior of the supervised plant. These concepts have been extended to several practical applications, including hierarchical Command, Control, Computer, Communication, Intelligence, Surveillance, and Reconnaissance (C^4ISR) systems [11]. Apparently, there have been no quantitative methods for evaluating the performance of supervisory controllers and establishing thresholds for their performance.

The concept of permissiveness has been used in DES control literature [1] [8] to facilitate qualitative comparison of DES controllers under the language controllability condition. Design of maximally permissive DES controllers has been proposed by several researchers based on different assumptions. However, maximal permissiveness does not imply best performance of the supervised plant from the perspective of achieving plant operational objectives. For example, in the travelling salesman problem, a maximally permissive supervisor may not yield the least expensive way of visiting the scheduled cities and returning to the starting point because no quantitative measure of performance is addressed in this type of supervisor design.

The above argument evinces the need for a signed real measure of regular languages, which can be used for quantitative evaluation and comparison of different supervisors for a physical plant, instead of relying on permissiveness as the (qualitative) performance index. Construction of the proposed language measure follows Myhill-Nerode Theorem [7], where a state-based partitioning of the (unsupervised) plant language yields equivalence classes of finite-length event strings. Each marked state is characterized by a signed real value that is chosen based on the designer's perception of the state's impact on the system performance. Conceptually similar to conditional probability, each event is assigned a cost based on the state at which it is generated. This procedure permits a string of events, terminating on a good (bad) marked state, to have a positive (negative) measure. A supervisor can be designed in this setting such that the supervisor attempts to eliminate as many bad strings as possible and retain as few good strings as possible. Different supervisors may achieve this goal in different ways and generate a partially ordered set of controlled languages. The language measure then creates a total ordering on the performance of the controlled languages, which provides a precise quantitative comparison of the controlled plant behavior under different supervisors. This feature is formally stated as follows:

> *Given that the relation $\subseteq$ induces a partial ordering on a set of controlled sublanguages $\{L(S^j/G), j = 1, \ldots, N\}$ of the plant language $L(G)$ under supervisors whose languages are $\{L(S^j), j = 1, \ldots, N\}$, the language measure μ induces a total ordering $\leq$ on $\{\mu(L(S^j/G))\}$. In other words, the range of the set function μ is totally ordered while its domain could be partially ordered.*

The above problem was first addressed by Wang and Ray [20] who proposed a signed measure of regular languages; an alternative approach was proposed by Ray and Phoha [15] who constructed a vector space of formal languages and defined a metric based on the total variation measure of the language. This chapter reviews these publications on language measure from the perspectives of discrete-event supervisory control within a unified framework and provides clarifications and extensions of the key concepts. Systematic procedures for computation of the language measure are developed in this chapter and they are illustrated with an engineering example.

The major objectives of this chapter are mathematically rigorous formulation and systematic construction of a real signed measure of regular languages based on the fundamental principles of real analysis and automata theory. This language measure quantifies sublanguages of a regular language and is readily applicable to analysis and synthesis of discrete-event supervisory control algorithms. Specifically, performance indices of DES supervisors can be defined in terms of the language measure.

The signed real measure for a DFSA, presented in this chapter, is constructed based on assignment of an event cost matrix and a characteristic vector. Two techniques for language measure computation have been recently reported. While the first technique [20] leads to a system of linear equations whose (closed form) solution yields the language measure vector, the second technique [15] is a recursive procedure with finite iterations. A sufficient condition for finiteness of the signed measure has been established in both cases.

In order to induce total ordering on the measure of different sublanguages of a plant language under different supervisors, it is implicit that same strings in different sublanguages must be assigned the same measure. This is accomplished by a quantitative tool that requires a systematic procedure to assign a characteristic vector and an event cost matrix. The clarifications and extensions presented in this chapter are intended to enhance development of a systematic analytical tool for synthesizing discrete-event supervisory control. For example, Fu et al. [5][4] have proposed unconstrained optimal control of regular languages where a state-based optimal control policy is obtained by selectively disabling controllable events to maximize the measure of the controlled plant language.

The chapter is organized in eight sections including the present introductory section and two appendices. Section 1.2 briefly describes the language measure and introduces the notations. Section 1.3 presents the procedure by which the performance of different supervisors can be compared based on a common quantitative tool. It also discusses two methods for computing language measure. Section 6.2.1 addresses issues regarding physical interpretation of the event cost used in the language measure. Section 1.5 presents a recursive algorithm for identification of the language parameters (i.e., elements of the event cost matrix). Section 1.6 illustrates the usage of the language measure for construction of metric spaces of formal languages and synthesis of optimal discrete-event supervisors. Section 1.7 presents an application of the language measure on the discrete-event model of a twin-engine unmanned aircraft [5]. The chapter is summarized and concluded in Section 1.8 along with recommendations for future research. Appendix A provides the pertinent mathematical background of measure theory as needed in the main body of the chapter. Appendix B establishes absolute convergence of the language measure.

1.2 Language Measure Concept

This section first introduces the signed real measure of regular languages, which is reported in [20] [15] [19].

Let $G_i \equiv \langle Q, \Sigma, \delta, q_i, Q_m \rangle$ be a trim (i.e., accessible and co-accessible) finite-state automaton model that represents the discrete-event dynamics of a physical plant, where $Q = \{q_k : k \in \mathcal{I}_Q\}$ is the set of states and $\mathcal{I}_Q \equiv \{1, 2, \cdots, n\}$ is the index set of states; the automaton starts with the initial state q_i; the alphabet of events is $\Sigma = \{\sigma_k : k \in \mathcal{I}_\Sigma\}$ with $\Sigma \bigcap \mathcal{I}_Q = \emptyset$, and $\mathcal{I}_\Sigma \equiv \{1, 2, \cdots, \ell\}$ is the index set of events; $\delta : Q \times \Sigma \to Q$ is the (possibly partial) function of state transitions; and $Q_m \equiv \{q_{m_1}, q_{m_2}, \cdots, q_{m_r}\} \subseteq Q$ is the set of marked (i.e., accepted) states with $q_{m_k} = q_j$ for some $j \in \mathcal{I}_Q$.

Let Σ^* be the Kleene closure of Σ, i.e., the set of all finite-length strings made of the events belonging to Σ as well as the empty string ϵ that is viewed as the identity of the monoid Σ^* under the operation of string concatenation, i.e., $\epsilon s = s = s\epsilon$. The extension $\hat{\delta} : Q \times \Sigma^* \to Q$ is defined recursively in the usual sense [7] [9] [14]. For discrete event supervisory control [14], the event alphabet Σ is partitioned into sets, Σ_c and $\Sigma - \Sigma_c$ of controllable and uncontrollable events, respectively, where each event in Σ_c and no event in $\Sigma - \Sigma_c$ can be disabled by the supervisor.

Definition 1.2.1 *The language $L(G_i)$ generated by a DFSA G_i initialized at the state $q_i \in Q$ is defined as:*

$$L(G_i) = \{s \in \Sigma^* \mid \hat{\delta}(q_i, s) \in Q\} \tag{1.1}$$

Since the state transition function δ is allowed to be a partial function, $L(G_i) \subseteq \Sigma^*$ following Definition 1.2.1; if δ is a total function, then the generated language $L(G_i) = \Sigma^*$.

Definition 1.2.2 *Given a DFSA plant model G_i, having the set of controllable events $\Sigma_c \subseteq \Sigma$, let S and $\widetilde{S}$ be two controllable supervisors (i.e., each of S and $\widetilde{S}$ is represented by an event disabling mapping $L(G_i) \to 2^{\Sigma_c}$). Let the languages of the plant supervised by S and $\widetilde{S}$ be denoted as $L(S/G_i)$ and $L(\widetilde{S}/G_i)$, respectively. Then, S is said to be less permissive (or more restrictive) than $\widetilde{S}$, denoted as $S \preceq \widetilde{S}$, if following condition holds:*

$$S \preceq \widetilde{S} \ \ if \ L(S/G_i) \subseteq L(\widetilde{S}/G_i) \tag{1.2}$$

In other words, S may disable a larger set of controllable events than $\widetilde{S}$ following the execution of any event string $s \in \Sigma^*$.

Definition 1.2.3 *The language $L_m(G_i)$ marked by a DFSA G_i initialized at the state $q_i \in Q$ is defined as:*

$$L_m(G_i) = \{s \in \Sigma^* \mid \hat{\delta}(q_i, s) \in Q_m\} \tag{1.3}$$

Definition 1.2.4 *For every $q_i, q_k \in Q$, let $L_{i,k}$ denote the set of all strings that, starting from the state q_i, terminate at the state q_k, i.e.,*

$$L_{i,k} = \{s \in \Sigma^* \mid \hat{\delta}(q_i, s) = q_k\} \tag{1.4}$$

In order to obtain a quantitative measure of the marked language, the set Q_m of marked states is partitioned into Q_m^+ and Q_m^-, i.e., $Q_m = Q_m^+ \cup Q_m^-$ and $Q_m^+ \cap Q_m^- = \emptyset$. The positive set Q_m^+ contains all *good* marked states that one would desire to reach, and the negative set Q_m^- contains all *bad* marked states that one would not want to terminate on, although it may not always be possible to completely avoid the *bad* states while attempting to reach the *good* states. From this perspective, each marked state is characterized by an assigned real value that is chosen based on the designer's perception of the state's impact on the system performance.

Definition 1.2.5 *The characteristic function* $\chi : Q \rightarrow [-1, 1]$ *assigns a signed real weight to a state-based sublanguage* $L_{i,j}$, *having each of its strings terminating on the same state* q_j, *and is defined as:*

$$\forall q_j \in Q, \quad \chi(q_j) \in \begin{cases} [-1,0), & q_j \in Q_m^- \\ \{0\}, & q_j \notin Q_m \\ (0,1], & q_j \in Q_m^+ \end{cases} \tag{1.5}$$

The state weighting vector, denoted by $\mathbf{X} = [\chi_1 \ \chi_2 \ \cdots \ \chi_n]^T$, *is called the* **X**-*vector, where* $\chi_j \equiv \chi(q_j)$. *That is, the* j^{th} *element* χ_j *of* **X**-*vector is the weight assigned to the corresponding state* q_j.

In general, the marked language $L_m(G_i)$ consists of both *good* and *bad* strings, which start from the initial state q_i, respectively lead to Q_m^+ and Q_m^-. Denoting the set difference operation by "−", any event string belonging to the language $L^0(G_i) \equiv L(G_i) - L_m(G_i)$ leads to one of the non-marked states belonging to $Q - Q_m$ and $L^0(G_i)$ does not contain any one of the *good* or *bad* strings. Partitioning Q_m into the positive set Q_m^+ and the negative set Q_m^- leads to partitioning of the marked language $L_m(G_i)$ into a positive language $L_m^+(G_i)$ and a negative language $L_m^-(G_i)$. Based on the equivalence classes defined in the Myhill-Nerode Theorem [7], the regular languages $L(G_i)$ and $L_m(G_i)$ can be expressed as:

$$L(G_i) = \cup_{k \in \mathcal{I}_Q} L_{i,k} \tag{1.6}$$

$$L_m(G_i) = L_m^+(G_i) \cup L_m^-(G_i) \tag{1.7}$$

where the sublanguage $L_{i,k} \subseteq L(G_i)$ is uniquely labeled by the state $q_k, k \in \mathcal{I}_Q$ and $L_{i,k} \cap L_{i,j} = \emptyset \ \forall k \neq j$; and $L_m^+(G_i) \equiv \cup_{q_k \in Q_m^+} L_{i,k}$ and $L_m^-(G_i) \equiv \cup_{q_k \in Q_m^-} L_{i,k}$ are *good* and *bad* sublanguages of $L_m(G_i)$, respectively. Then, the null sublanguage $L^0(G_i) = \cup_{q_k \notin Q_m} L_{i,k}$ and $L(G_i) = L^0(G_i) \cup L_m^+(G_i) \cup L_m^-(G_i)$.

Now a signed real measure is constructed as $\mu^i : 2^{L(G_i)} \rightarrow \mathbf{R} \equiv (-\infty, \infty)$ on the σ−algebra $M = 2^{L(G_i)}$. (Appendix A provides details of measure-theoretic definitions and results.) With this choice of σ−algebra, every singleton set made of an event string $s \in L(G_i)$ is a measurable set, which allows its quantitative evaluation based on the above state-based decomposition of $L(G_i)$ into null (i.e., L^0), positive (i.e., L_m^+), and negative (i.e., L_m^-) sublanguages.

Conceptually similar to the conditional probability, each event is assigned a cost based on the state at which it is generated.

Definition 1.2.6 *The event cost of the DFSA* G_i *is defined as a (possibly partial) function* $\tilde{\pi} : \Sigma^* \times Q \rightarrow [0, 1]$ *such that* $\forall q_i \in Q$, $\forall \sigma_j \in \Sigma$, $\forall s \in \Sigma^*$,

$$\tilde{\pi}[\sigma_j, q_i] = 0 \text{ if } \delta(q_i, \sigma_j) \text{ is undefined; } \tilde{\pi}[\epsilon, q_i] = 1;$$

$$\tilde{\pi}[\sigma_j, q_i] \equiv \tilde{\pi}_{ij} \in [0, 1); \quad \sum_{j \in \mathcal{I}_\Sigma} \tilde{\pi}_{ij} < 1; \tag{1.8}$$

$$\tilde{\pi}[\sigma_j s, q_i] = \tilde{\pi}[\sigma_j, q_i] \ \tilde{\pi}[s, \delta(q_i, \sigma_j)].$$

A simple application of the induction principle to the last part of Definition 1.2.6 shows $\tilde{\pi}[st, q_j] = \tilde{\pi}[s, q_j]\tilde{\pi}[t, \delta^*(q_j, s)]$. The condition $\sum_{k \in \mathcal{I}_Q} \tilde{\pi}_{jk} < 1$ provides a sufficient condition for the existence of the real signed measure as discussed in Section 1.3. Additional comments on the physical interpretation of the event cost are provided in Section 6.2.1.

Consequently, the $n \times \ell$ event cost matrix is defined as:

$$\widetilde{\Pi} = \begin{bmatrix} \tilde{\pi}_{11} & \tilde{\pi}_{12} & \cdots & \tilde{\pi}_{1\ell} \\ \tilde{\pi}_{21} & \tilde{\pi}_{22} & \cdots & \tilde{\pi}_{2\ell} \\ \vdots & \vdots & \ddots & \vdots \\ \tilde{\pi}_{n1} & \tilde{\pi}_{n2} & \cdots & \tilde{\pi}_{n\ell} \end{bmatrix} \tag{1.9}$$

Definition 1.2.7 *The state transition cost, $\pi : Q \times Q \to [0, 1)$, of the DFSA G_i is defined as follows:*
$\forall i, j \in \mathcal{I}_Q$,

$$\pi_{ij} = \begin{cases} \sum_{\sigma \in \Sigma} \tilde{\pi}[\sigma, q_i], & \text{if } \delta(q_i, \sigma) = q_j \\ 0 & \text{if } \{\delta(q_i, \sigma) = q_j\} = \emptyset. \end{cases} \tag{1.10}$$

Consequently, the $n \times n$ state transition cost matrix is defined as:

$$\Pi = \begin{bmatrix} \pi_{11} & \pi_{12} & \cdots & \pi_{1n} \\ \pi_{21} & \pi_{22} & \cdots & \pi_{2n} \\ \vdots & \vdots & \ddots & \vdots \\ \pi_{n1} & \pi_{n2} & \cdots & \pi_{nn} \end{bmatrix} \tag{1.11}$$

and is referred to as the Π-matrix in the sequel.

Definition 1.2.8 *For a given DFSA G_i, the signed real measure of every singleton string set $\{s\} \in L_{i,j} \subseteq L(G_i)$ is defined as: $\mu^i(\{s\}) \equiv \tilde{\pi}(s, q_i)\chi_j$ implying that*

$$\forall s \in L_{i,j}, \qquad \mu^i(\{s\}) \begin{cases} = 0, & q_j \notin Q_m \\ > 0, & q_j \in Q_m^+ \\ < 0, & q_j \in Q_m^- \end{cases} \tag{1.12}$$

Thus an event string terminating on a *good* (*bad*) marked state has a *positive* (*negative*) measure and one terminating on a non-marked state has *zero* measure. It follows from Definition 1.2.8 that the signed measure of the sublanguage $L_{i,j} \subseteq L(G_i)$ of all events, starting at q_i and terminating at q_j, is:

$$\mu^i(L_{i,j}) = \left(\sum_{s \in L_{i,j}} \tilde{\pi}[s, q_i] \right) \chi_j \tag{1.13}$$

Definition 1.2.9 *Given a DFSA $G_i \equiv \langle Q, \Sigma, \delta, q_i, Q_m \rangle$ the cost ν^i of a sublanguage $K \subseteq L(G_i)$ is defined as the sum of the event cost $\tilde{\pi}$ of individual strings belonging to K:*

$$\nu^i(K) = \sum_{s \in K} \tilde{\pi}[s, q_i] \qquad (1.14)$$

Definition 1.2.10 *The signed real measure of the language of a DFSA G_i initialized at a state $q_i \in Q$, is defined as:*

$$\mu_i \equiv \mu^i(L(G_i)) = \sum_{j \in \mathcal{I}_Q} \mu^i(L_{i,j}) \qquad (1.15)$$

The language measure vector, denoted as $\mu = [\mu_1 \ \mu_2 \ \cdots \ \mu_n]^T$, is called the μ-vector.

Remark 1.2.1 $\mu^i(L_m(G_i)) = \mu_i \ \forall i \in \mathcal{I}_Q$ *because* $\chi_k = 0 \ \forall q_k \in Q - Q_m$.

It follows from Definition 1.2.10 that $\mu^i(L_{i,j}) = \nu^i(L_{i,j})\chi_j$. Under the condition of $\sum_k \tilde{\pi}_{jk} < 1$ in Definition 1.2.6, convergence of the signed real language measure μ^i has been proved in [15] [20]. The total variation measure $|\mu^i|$ of μ^i has also been shown to be finite for every $i \in \mathcal{I}_Q$ [15].

In the above setting, the role of the language measure in DES control synthesis is explained below:

A discrete-event non-marking supervisor S restricts the marked behavior of an uncontrolled (i.e., unsupervised) plant G_i such that $L_m(S/G_i) \subseteq L_m(G_i)$. The uncontrolled marked language $L_m(G_i)$ consists of good strings leading to Q_m^+ and bad strings leading to Q_m^-. A controlled language $L_m(S/G_i)$ based on a given specification of the supervisor S may disable some of the bad strings and keep some of the good strings enabled. Different supervisors $S_j : j \in \{1, 2, \ldots, n_s\}$ for a DFSA G_i achieve this goal in different ways and generate a partially ordered set of controlled sublanguages $\{L_m(S_j/G_i) : j \in \{1, 2, \ldots, n_s\}\}$. The real signed measure μ^i provides a precise quantitative comparison of the controlled plant behavior under different supervisors because the set $\{\mu^i(L_m(S_j/G_i)) : j \in \{1, 2, \ldots, n_s\}\}$ is totally ordered.

In order to realize the above goal, the performance of different supervisors has to be evaluated based on a common quantitative tool. Let $G \equiv \langle Q_1, \Sigma, \delta_1, q_{11}, Q_{m1} \rangle$ denote the uncontrolled plant and $S \equiv \langle Q_2, \Sigma, \delta_2, q_{21}, Q_{m2} \rangle$ denote the supervisor with corresponding marked languages $L_m(G)$ and $L_m(S)$, respectively. Then $L_m(G)$ denotes the uncontrolled plant language, $L_m(S)$ is language of control specification and $L_m(G) \cap L_m(S)$ is the controlled sublanguage under the supervisor S.

Let $C \equiv \langle Q, \Sigma, \delta, q_1, Q_m \rangle$ where, $Q = Q_1 \times Q_2$, $q_1 = (q_{11}, q_{21})$, $Q_m = \{(p, q) | p \in Q_{m1} \text{ and } q \in Q_{m2}\}$ and the transition function δ is defined by the formula:

$$\delta((p, q), \sigma) = (\delta_1(p, \sigma), \delta_2(q, \sigma)), \qquad (1.16)$$

$(\forall p \in Q_1, \ q \in Q_2,$ and $\sigma \in \Sigma)$. Then, based on the established results from automata theory [9] and supervisory control [14], it is concluded that the automaton C accepts the language $L_m(G) \cap L_m(S)$.

It follows from the above discussion that the extension $\hat{\delta}$ satisfies: $\forall s \in \Sigma^*$,

$$\hat{\delta}((p,q),s) = (\hat{\delta}_1(p,s), \hat{\delta}_2(q,s)) \tag{1.17}$$

Also $L(G)$ is partitioned by $L_{11,1j}$, $1 \leq j \leq n_1$ where $|Q_1| = n_1$. With the above construction, each of $L_{11,1j}$ is further partitioned by $L_{11,1j} \cap L_{21,2k}$, $1 \leq k \leq n_2$, where $|Q_2| = n_2$. Thus, for any $q_{1j} \in Q_{m1}$, the set of strings, which is retained in $L_m(G) \cap L_m(S)$, is given by $L_{11,1j} \cap (\cup_{q_{2k} \in Q_{m2}} L_{21,2k})$. A supervisor which retains maximum possible strings corresponding to $q_{1j} \in Q_{m1}^+$ while discarding as many strings as possible corresponding to $q_{1j} \in Q_{m1}^-$ would give a higher measure and hence a better performance. The construction above shows immediately how the event cost and characteristic function assigned to the uncontrolled plant can be used as a quantitative tool with which the performance of different supervisors can be evaluated and compared. The following procedure indicates how this can be accomplished explicitly.

Definition 1.2.11 *Let G, S and C be defined as above. Let G represent the unsupervised plant and $\tilde{\pi}_1$ be the event cost and χ_1 be the characteristic function. Then for the DFSA C which represents the controlled sublanguage, $\tilde{\pi}$ is defined as:*

$$\tilde{\pi}[\sigma, (q_{1i}, q_{2j})] = \tilde{\pi}_1[\sigma, q_{1i}] \tag{1.18}$$

$\forall \sigma \in \Sigma$ and $\forall i, j$ s.t. $1 \leq i \leq n_1$, $1 \leq j \leq n_2$.
The χ-vector is defined as:

$$\chi((q_{1i}, q_{2j})) = \chi_1(q_{1i}) \mathcal{J}(q_{2j}) \tag{1.19}$$

where $\mathcal{J}(\cdot)$ is the indicator function defined as:

$$\mathcal{J}(q) = \begin{cases} 1 & q \in Q_{m2} \\ 0 & q \notin Q_{m2} \end{cases} \tag{1.20}$$

Let $s \in L((q_{11}, q_{21}), (q_{1i}, q_{2j}))$. Since $q_{1i} = \hat{\delta}(q_{11}, s)$, by Definition 1.2.8, $\mu(\{s\}) = \tilde{\pi}_1[s, q_{11}] \chi_1(q_{1i})$ for the unsupervised (i.e., open loop) plant. For the supervised (i.e., closed loop) plant, the corresponding measure is: $\mu(\{s\}) = \tilde{\pi}[s, (q_{11}, q_{21})] \chi((q_{1i}, q_{2j})) = \tilde{\pi}_1[s, q_{11}] \chi_1(q_{1i}) \mathcal{J}(q_{2j})$ by Equations (1.18) and (1.19). In other words, if no event in the string s is disabled by the supervisor, then $\mu^i(\{s\})$ remains unchanged; otherwise, $\mu^i(\{s\}) = 0$. Thus, Definition 1.2.11 guarantees that the same strings in different controlled sublanguages of a plant language $L(G_i)$ are assigned the same measure. Hence, the performance of different supervisors can be compared with a common quantitative tool.

Finally to conclude, it should be noted that while the domain (i.e., $2^{L(G_i)}$) of the language measure μ^i is partially ordered, its range which is a subset of

R becomes totally ordered. The set $L(G_i)$ with the σ-algebra, $2^{L(G_i)}$, forms a measurable space. In principle, any measure μ can be defined on this measurable space to form a measure space (i.e., the triple $\langle L(G_i), 2^{L(G_i)}, \mu^i \rangle$). The choice of the signed language measure as given by Definitions 1.2.8 and 1.2.10, has been the motivated by the fact that it may serve as a performance measure and hence should have a physical significance in the DES controller synthesis. Moreover, defining the measure in this way also leads to simple computational procedures as discussed in the next section and further elaborated later in Section 6.2.1.

1.3 Language Measure Computation

Various methods of obtaining regular expressions for DFSAs are reported in Hopcroft [7], Martin [9], and Drobot [3]. While computing the measure of a given DFSA, the same event may have different significance when emanating from different states. This requires assigning (possibly) different costs to the same event defined on different states. Therefore, it is necessary to obtain a regular expression which explicitly yields the state-based event sequences. In order to compute the language measure, it is convenient to transform the procedures of evaluating regular expression from symbolic equations to algebraic ones. The following two methods [20] [15] are presented, in detail, for language measure computation.

1.3.1 Method I: Closed Form Solution

This section presents a closed-form method to compute the language measure via inversion of a square operator.

Definition 1.3.1 *Let $L_i \equiv L(G_i), i \in \mathcal{I}_Q$, denote the regular expression representing the language of a DFSA $G_i = \langle Q, \Sigma, \delta, q_i, Q_m \rangle$ where q_i is the initial state.*

Definition 1.3.2 *Let σ_j^k denote the set of event(s) $\sigma \in \Sigma$ that is defined on the state q_j and leads to the state $q_k \in Q$, where $j, k \in \mathcal{I}_Q$, i.e., $\delta(q_j, \sigma) = q_k, \forall \sigma \in \sigma_j^k \subseteq \Sigma$.*

Then, given a DFSA $G_i = \langle Q, \Sigma, \delta, q_i, Q_m \rangle$ with $|Q| = n$, the procedure to obtain the system equation by a set of regular expressions L_i of the language $L(G_i), i \in \mathcal{I}_Q$, as follows:

$$\forall q_i \in Q, \qquad L_i = \sum_{j \in \mathcal{I}_Q} R_{i,j} + \mathcal{E}_i, \qquad i \in \mathcal{I}_Q \qquad (1.21)$$

where the operator $\sum$ indicates the sum of regular expressions (equivalently, union of regular languages); and $\forall i$, $R_{i,j}$ and $\mathcal{E}_i$ are defined as:

1. If $\exists\ \sigma \in \Sigma$, such that $\delta(q_i, \sigma) = q_j \in Q, j \in \{1, \cdots, n\}$, then $R_{i,j} = \sigma_i^j L_j$, otherwise, $R_{i,j} = \emptyset$.
2. If $q_i \in Q_m$, $\mathcal{E}_i = \epsilon$, otherwise, $\mathcal{E}_i = \emptyset$.

The set of symbolic equations may be written as:

$$L_i = \sum_j \sigma_i^j L_j + \mathcal{E}_i \tag{1.22}$$

The above system of symbolic equations can be solved using a result given below, which is illustrated through an example.

Lemma 1.3.1 *Let u, v be two known regular expressions and r be an unknown regular expression that satisfies the following algebraic identity:*

$$r = ur + v \tag{1.23}$$

Then, the following relations are true:

(1) $r = u^*v$ is a solution to Equation (1.23).
(2) If $\epsilon \notin u$, then $r = u^*v$ is the unique solution to Equation (1.23).

Proof. The proof of Lemma 1.3.1, which is also known as Arden's relation, is given in [3] [15].

Example 1. In this example, shown in Figure 1.1, the alphabet is $\Sigma = \{a, b\}$; the set of states is $Q = \{1, 2, 3\}$; the initial state is 1; and the sole marked state is 2. Let the set of linear algebraic equations representing the transitions at each state of the DFSA be:

$$\begin{cases} L_1 = a_1^1 L_1 + b_1^2 L_2 + \mathcal{E}_1 = a_1^1 L_1 + b_1^2 L_2 \\ L_2 = a_2^1 L_1 + b_2^3 L_3 + \mathcal{E}_2 = a_2^1 L_1 + b_2^3 L_3 + \epsilon \\ L_3 = a_3^1 L_1 + b_3^2 L_2 + \mathcal{E}_3 = a_3^1 L_1 + b_3^2 L_2 \end{cases} \tag{1.24}$$

where the 'forcing' term ϵ is introduced on the right side of the ith equation whenever $q_i \in Q_m, i \in \mathcal{I}_Q$. By application of Lemma 1.3.1, the regular expression for the language $L(G_1)$ is:

$$L(G_1) \equiv L_1 = (a_1^1)^* b_1^2 (a_2^1 (a_1^1)^* b_1^2 + b_2^3 a_3^1 (a_1^1)^* b_1^2 + b_2^3 b_3^2)^*$$

Instead of obtaining regular expressions, the language measure can be directly computed by transforming this set of equations into a system of linear equations based on the following result.

Theorem 1.3.1 *Following Definition 1.2.10, the language measure of the symbolic equations 1.22 is given by*

$$\mu_i = \sum_j \pi_{ij} \mu_j + \chi_i \tag{1.25}$$

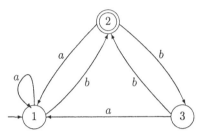

Fig. 1.1. Finite State Machine for Example 1

Proof. Following Equation (1.21) and Definition 1.2.5:

$$\forall i \in \mathcal{I}_Q, \qquad \mu^i(\mathcal{E}_i) = \begin{cases} \chi_i & \text{if } \mathcal{E}_i = \epsilon \\ 0 & \text{otherwise} \end{cases} \qquad (1.26)$$

Therefore, each element of the vector $\mathbf{X} = [\chi_1 \; \chi_2 \; \cdots \; \chi_n]^T$ is the forcing function in Equations (1.22) and (1.25). Starting from the state q_i, the measure of the language $L_i \equiv L_m(G_i)$ (see Definition 1.3.1)

$$\mu_i = \mu^i(L_i) = \mu^i \left(\sum_j \sigma_i^j L_j + \mathcal{E}_i \right)$$

$$= \mu^i \left(\sum_j \sigma_i^j L_j \right) + \mu^i(\mathcal{E}_i)$$

$$= \sum_j \mu^i(\sigma_i^j L_j) + \mu^i(\mathcal{E}_i)$$

$$= \sum_j \pi(\sigma_i^j)\mu^j(L_j) + \mu^i(\mathcal{E}_i)$$

$$= \sum_j \pi_{ij}\mu_j + \chi_i$$

The third equality in the above derivation follows from the fact that $\mathcal{E}_i \cap \sigma_i^j L_j = \emptyset$. It is also true that

$$\forall j \neq k, \qquad \sigma_i^j L_j \cap \sigma_i^k L_k = \emptyset \qquad (1.27)$$

since each string in $\sigma_i^j L_j$ starts with an event in σ_i^j while each string in $\sigma_i^k L_k$ starts from an event in σ_i^k and $\sigma_i^j \cap \sigma_i^k = \emptyset \; \forall j \neq k$ because G_i is a DFSA. This justifies the fourth equality. The fifth equality follows from Definition 1.2.9 and the fact that $\mu^i(L_{i,j}) = \nu^i(L_{i,j})\chi(q_j)$; therefore, by Definitions 1.2.7 and 1.3.2, $\mu^i(\sigma_i^j L_j) = \pi[q_i, q_j] \, \mu^j(L_j) = \pi_{ij}\mu_j$.

In vector notation, Equation (1.25) in Theorem 1.3.1 is expressed as:

$$\mu = \Pi\mu + X$$

whose solution is given by:

$$\mu = (I - \Pi)^{-1}X \qquad (1.28)$$

provided that the matrix $I - \Pi$ is invertible. The following important result guarantees the existence of μ:

Theorem 1.3.2 *Given DFSAs $G_i \equiv \langle Q, \Sigma, \delta, q_i, Q_m \rangle$, $1 \leq i \leq n$ with the state transition cost matrix Π. Then the matrix $(I-\Pi)$ is an invertible bounded linear operator and $\mu \in \mathbf{R}^n$.*

Proof. It follows from Definitions 1.2.6 and 1.2.7 that the induced infinity norm $||\Pi||_\infty \equiv \max_i \sum_j \pi_{ij} = 1 - \theta$ where $\theta \in (0,1)$. Then $(I - \Pi)$ is invertible and is a bounded linear operator and the induced infinity norm $||(I - \Pi)^{-1}||_\infty \leq \theta^{-1}$ [10]. Then, it follows immediately from Equation (1.28) that $\mu \in \mathbf{R}^n$.

Corollary 1.3.1 (to Theorem 1.3.2) *The language measure vector μ is bounded as: $||\mu||_\infty \leq \theta^{-1}$ where $\theta \equiv (1 - ||\Pi||_\infty)$.*

Proof. The proof follows by applying the norm inequality property and Theorem 1.3.2 to Equation (1.28) and the fact that the max norm $||X||_\infty \leq 1$ by Definition 1.2.5.

Alternatively, sufficient conditions for convergence of μ can be obtained based on the properties of nonnegative matrices that are given in Appendix B. Therefore, Definitions 1.2.6 and 1.2.7 provide a sufficient condition for the language measure μ of the DFSA G_i to be finite. A closed-form algorithm to compute a language measure based on the above procedure is presented below.

Algorithm 1.3.1 Closed-form computation of the language measure

(1) For a given $G_i \equiv \langle Q, \Sigma, \delta, q_i, Q_m \rangle$, specify the characteristic vector X (see Definition 1.2.5) and determine the event cost matrix $\widetilde{\Pi}$ (see Definition 1.2.6) via experimentation or simulation, as described in Section 1.5).

(2) Generate the Π-matrix (Definition 1.2.7).

(3) Compute the language measure vector $\mu \leftarrow (I - \Pi)^{-1}X$ using Gaussian elimination.

(4) Obtain μ_i, the ith element of μ-vector, which is the measure of the generated language of the DFSA G_i.

The jth element of the ith row of the $(\mathbf{I}-\mathbf{\Pi})^{-1}$ matrix, denoted as ν_i^j, is the language measure of the DFSA with the same state transition function δ as G_i and having the following properties: (i) the initial state is q_i; (ii) q_j is the only marked state; and (iii) the χ-value of q_j is equal to 1. Thus, $\mu_i \equiv \mu(L_m(G_i))$ is given by $\mu_i = \sum_j \nu_i^j \chi_j$. Numerical evaluation of the language measure of the automaton G_i requires Gaussian elimination of the single variable μ_i involving the real invertible matrix $(\mathbf{I} - \mathbf{\Pi})$. Therefore, the computational complexity of the language measure algorithm is polynomial in the number of states.

1.3.2 Method II: Recursive Solution

This section presents a second method to compute the language measure using a recursive procedure based on concept of Kleene's theorem [9] which shows that a language accepted by a DFSA is regular. It also yields an algorithm to recursively construct the regular expression of its marked language instead of the the closed form solution in Method I.

Definition 1.3.3 *Given q_i, $q_k \in Q$, a non-empty string p of events (i.e. $p \neq \epsilon$) starting from q_i and terminating at q_k is called a path. A path p from q_i to q_k is said to pass through q_j if $\exists s \neq \epsilon$ and $t \neq \epsilon$ such that $p = st$; $\hat{\delta}(q_i, s) = q_j$ and $\hat{\delta}(q_j, t) = q_k$.*

Definition 1.3.4 *A path language p_{ik}^j is defined to be the set of all paths from q_i to q_k, which do not pass through any state q_r for $r > j$. The path language p_{ik} is defined to be the set of all paths from q_i to q_k. Thus, the language $L_{i,k}$ is obtained in terms of the path language p_{ik} as:*

$$L_{i,k} = \begin{cases} p_{ii} \cup \{\epsilon\}, & if \quad k = i \\ p_{ik}, & if \quad k \neq i \end{cases}$$

$$\Rightarrow \nu(L_{i,k}) = \begin{cases} \nu(p_{ii}) + 1, & if \quad k = i \\ \nu(p_{ik}), & if \quad k \neq i \end{cases}$$

Every path language p_{ik}^j is a regular language and subset of $L(G_i)$. As shown in [15], following recursive relation holds for $0 \leq j \leq n - 1$:

Theorem 1.3.3 *Given a $G_i \equiv \langle Q, \Sigma, \delta, q_i, Q_m \rangle$, the following recursive relation holds for $1 \leq j \leq n - 1$*

$$p_{lk}^0 = \{\sigma \in \Sigma : \delta(q_l, \sigma) = q_k\}$$
$$p_{lk}^{j+1} = p_{lk}^j \bigcup p_{l,j+1}^j \left(p_{j+1,j+1}^j \right)^* p_{j+1,k}^j \tag{1.29}$$

Proof. Since the states are numbered form 1 to n in increasing order, $p_{lk}^0 = \{\sigma \in \Sigma : \delta(q_l, \sigma) = q_k\}$ follows directly form the state transition map $\delta : Q \times \Sigma \to Q$ and Definition 1.3.4. Given $p_{lk}^j \subseteq p_{lk}^{j+1}$, let us consider the set $p_{lk}^{j+1} - p_{lk}^{j}$ in which each string passes through q_{j+1} in the path from q_l to q_k and no string must pass through q_m for $m > (j+1)$. Then, it follows that $p_{lk}^{j+1} - p_{lk}^{j} = p_{l,j+1}^{j}p_{j+1,k}^{j+1}$ where $p_{j+1,k}^{j+1}$ can be expanded as: $p_{j+1,k}^{j+1} = \left(p_{j+1,j+1}^{j}p_{j+1,k}^{j+1}\right)\bigcup p_{j+1,k}^{j}$ that has a unique solution by Theorem 1.3.1 because $\epsilon \notin p_{j+1,j+1}^{j}$ based on Definition 1.3.4. Therefore,

$$p_{lk}^{j+1} = p_{lk}^{j} \bigcup p_{l,j+1}^{j}\left(p_{j+1,j+1}^{j}\right)^{*}p_{j+1,k}^{j}$$

Based on the three lemmas proved below, the above relations can be transformed into an algebraic equation conceptually similar to Theorem 1.3.1 in Method I. Along with the procedure to compute the language measure it is established that, $\forall i \in \mathcal{I}_Q$, $\sum_{j=1}^{n} \pi_{ij} < 1$ is a sufficient condition for finiteness of μ.

Lemma 1.3.2 $\nu\left(\left(p_{kk}^{0}\right)^{*}\left(\bigcup_{j\neq k}p_{kj}^{0}\right)\right) \in [0,1)$

Proof. Following Definition 1.2.6, $\nu(p_{kk}^{0}) \in [0,1)$. Therefore by convergence of geometric series,

$$\nu\left(\left(p_{kk}^{0}\right)^{*}\left(\bigcup_{j\neq k}p_{kj}^{0}\right)\right) = \frac{\sum_{j\neq k}\nu\left(p_{kj}^{0}\right)}{1 - \nu\left(p_{kk}^{0}\right)} \in [0,1)$$

because $\sum_{j}\nu\left(p_{kj}^{0}\right) < 1 \Rightarrow \sum_{j\neq k}\nu\left(p_{kj}^{0}\right) < 1 - \nu(p_{kk}^{0})$.

Lemma 1.3.3 $\nu\left(p_{j+1,j+1}^{j}\right) \in [0,1)$.

Proof. The path $p_{j+1,j+1}^{j}$ may contain at most j loops, one around each of the states $q_1, q_2 \cdots, q_j$. If the path $p_{j+1,j+1}^{j}$ does not contain any loop, then $\nu\left(p_{j+1,j+1}^{j}\right) \in [0,1)$ because $\forall s \in p_{j+1,j+1}^{j}$, $\nu(s) < 1$ and each of s originates at state $j+1$. Next suppose there is a loop around q_l and that does not contain any other loop; this loop must be followed by one or more events σ_k generated at q_l and leading to some other states q_m where $m \in \{1, \cdots, j+1\}$ and $m \neq l$. By Lemma 1.3.2, $\nu\left(p_{j+1,j+1}^{j}\right) \in [0,1)$. Proof follows by starting from the innermost loop and ending with all loops at q_j

Lemma 1.3.4

$$\nu\left(\left(p_{j+1,j+1}^{j}\right)^{*}\right) = \frac{1}{1 - \nu\left(p_{j+1,j+1}^{j}\right)} \in [1,\infty) \qquad (1.30)$$

Proof. Since $\nu\left(p_{j+1,j+1}^j\right) \in [0,1)$ from Lemma 1.3.3, it follows that

$$\nu\left(\left(p_{j+1,j+1}^j\right)^*\right) = \frac{1}{1 - \nu\left(p_{j+1,j+1}^j\right)} \in [1, \infty)$$

Finally, the main result of this section is stated as the following theorem.

Theorem 1.3.4 *Given a DFSA* $G_i \equiv \langle Q, \Sigma, \delta, q_i, Q_m \rangle$ *the following recursive result holds for* $0 \le j \le n-1$*:*

$$\nu\left(p_{lk}^{j+1}\right) = \nu\left(p_{lk}^j\right) + \frac{\nu\left(p_{l,j+1}^j\right)\nu\left(p_{j+1,k}^j\right)}{1 - \nu\left(p_{j+1,j+1}^j\right)} \tag{1.31}$$

Proof.

$$\begin{aligned}
\nu\left(p_{lk}^{j+1}\right) &= \nu\left(p_{lk}^j \bigcup p_{l,j+1}^j \left(p_{j+1,j+1}^j\right)^* p_{j+1,k}^j\right) \\
&= \nu\left(p_{lk}^j\right) + \nu\left(p_{l,j+1}^j \left(p_{j+1,j+1}^j\right)^* p_{j+1,k}^j\right) \\
&= \nu\left(p_{lk}^j\right) + \nu\left(p_{l,j+1}^j\right)\nu\left(\left(p_{j+1,j+1}^j\right)^*\right)\nu\left(p_{j+1,k}^j\right) \\
&= \nu\left(p_{lk}^j\right) + \frac{\nu\left(p_{l,j+1}^j\right)\nu\left(p_{j+1,k}^j\right)}{1 - \nu\left(p_{j+1,j+1}^j\right)}
\end{aligned}$$

where the second step follows from fact that $p_{lk}^j \bigcap p_{l,j+1}^j \left(p_{j+1,j+1}^j\right)^* p_{j+1,k}^j = \emptyset$. The third step follows from Definition 1.2.9 and the last step is a result of Lemma 1.3.4.

Based on the above result, a recursive algorithm to compute a language measure is presented below.

Algorithm 1.3.2 Recursive computation of the language measure

(1) *For a given* $G_i \equiv \langle Q, \Sigma, \delta, q_i, Q_m \rangle$*, specify the characteristic vector* **X** *(see Definition 1.2.5) and determine the event cost matrix* $\widetilde{\Pi}$ *(see Definition 1.2.6)via experimentation or simulation, as described in Section 1.5).*
(2) *Compute the* **Π**-matrix *(Definition 1.2.7)*
(3) $\nu(p_{lk}^0) \longleftarrow \pi_{lk}$ *for* $1 \le l, k \le n$
(4) *for j=0 to n-1*
    for l=1 to n
        for k=1 to n
        $\nu(p_{lk}^{j+1}) = \nu(p_{lk}^j) + \frac{\nu(p_{l,j+1}^j)\nu(p_{j+1,k}^j)}{1-\nu(p_{j+1,j+1}^j)}$
        end
    end
  end

(5) Calculate $\nu(L_{i,k})$ from $\nu(p_{ik})$ using Definition 1.3.4

(6) $\mu_i \longleftarrow \sum_{q_j \in Q_m} \nu(L_{i,j})\chi_j$ is the measure of marked language of DFSA, G_i

Since there are only three *for* loops, the computational complexity of the above algorithm is polynomial in the number of DFSA states, same as that of the algorithm 1.3.1 in Method I.

1.4 Event Cost: A Probabilistic Interpretation

The signed real measure (Definition 1.2.10) of a regular language is based on the assignment of the characteristic vector and the event cost matrix. As stated earlier, the characteristic vector is chosen by the designer based on his/her perception of the individual state's impact on the system performance. On the other hand, the event cost is an intrinsic property of the plant. The event cost $\tilde{\pi}_{jk}$ is conceptually similar to the state-based conditional probability of Markov Chains, except for the fact that it is not allowed to satisfy the equality condition $\sum_k \tilde{\pi}_{jk} = 1$. (Note that $\sum_k \tilde{\pi}_{jk} < 1$ is a requirement for convergence of the language measure.) The rationale for this strict inequality is explained below.

Since the plant model is an inexact representation of the physical plant, there exist unmodelled dynamics to account for. This can manifest itself either as unmodelled events that may occur at each state or as unaccounted states in the model. Let Σ_j^u denote the set of all unmodelled events at state q_j of the DFSA $G_i \equiv \langle Q, \Sigma, \delta, q_i, Q_m \rangle$. Creating a new unmarked absorbing state q_{n+1}, called the dump state [14], and extending the transition function δ to $\delta_{ext} : (Q \cup \{q_{n+1}\}) \times (\Sigma \cup_j \Sigma_j^u) \to (Q \cup \{q_{n+1}\})$, it follows that:

$$\delta_{ext}(q_j, \sigma) = \begin{cases} \delta_{ext}(q_j, \sigma), & \text{if } q_j \in Q \text{ and } \sigma \in \Sigma \\ q_{n+1}, & \text{if } q_j \in Q \text{ and } \sigma \in \Sigma_j^u \\ q_{n+1}, & \text{if } j = n+1 \text{ and } \sigma \in \Sigma \cup \Sigma_j^u \end{cases} \quad (1.32)$$

Therefore the residue $\theta_j = 1 - \sum_k \tilde{\pi}_{jk}$ denotes the probability of the set of unmoderated events Σ_j^u conditioned on the state j. The $\mathbf{\Pi}$ matrix can be similarly augmented to obtain a stochastic matrix $\mathbf{\Pi}_{aug}$ as follows:

$$\mathbf{\Pi}_{aug} = \begin{bmatrix} \pi_{11} & \pi_{12} & \cdots & \pi_{1n} & \theta_1 \\ \pi_{21} & \pi_{22} & \cdots & \pi_{2n} & \theta_2 \\ \vdots & \vdots & \ddots & \vdots & \vdots \\ \pi_{n1} & \pi_{n2} & \cdots & \pi_{nn} & \theta_n \\ 0 & 0 & \cdots & 0 & 1 \end{bmatrix} \quad (1.33)$$

Since the dump state q_{n+1} is not marked, its characteristic value $\chi_{n+1} \equiv \chi(q_{n+1}) = 0$. The characteristic vector then augments to $\mathbf{X}_{aug} = [\mathbf{X}^T \ 0]^T$.

With these extensions the language measure vector $\boldsymbol{\mu_{aug}} \equiv [\mu_1\,\mu_2\,\cdots\,\mu_n\,\mu_{n+1}]^T$ $=[\boldsymbol{\mu}^T\,\mu_{n+1}]^T$ of the augmented DFSA $G_{aug} \equiv \langle Q \cup \{q_{n+1}\}, \Sigma \cup_j \Sigma_j^u, \delta_{ext}, q_i, Q_m \rangle$ can be expressed as:

$$\boldsymbol{\mu_{aug}} \equiv \begin{pmatrix} \boldsymbol{\mu} \\ \mu_{n+1} \end{pmatrix} = \begin{pmatrix} \boldsymbol{\Pi}\boldsymbol{\mu} + \mu_{n+1}\,[\theta_1 \cdots \theta_n]^T \\ \mu_{n+1} \end{pmatrix} + \begin{pmatrix} \mathbf{X} \\ 0 \end{pmatrix} \qquad (1.34)$$

Since $\chi(q_{n+1}) = 0$ and all transitions from the absorbing state q_{n+1} lead to itself, $\mu_{n+1} = \mu(L_m(G_{n+1})) = 0$. Hence, Equation (6.11) reduces to that for the original plant G_i. Thus, the event cost can be interpreted as conditional probability, where the residue $\theta_j = 1 - \sum_k \tilde{\pi}_{jk} > 0$ accounts for the probability of all unmodelled events emanating from the state q_j. With this interpretation of event cost, $\tilde{\pi}[s, q_i]$ (Definition 1.2.6) denotes the probability of occurrence of the event string s in the plant model G_i starting at state q_i and terminating at state $\delta^*(s, q_i)$. Hence, $\nu(L_{i,j})$ (Definition 1.2.9), which is a non-negative real number, is directly related to the total probability that state q_i would be reached as the plant operates. (Note that $\nu(L_{i,j}) > 1$ is possible if $L_{i,j}$ contains multiple strings.) The language measure $\mu_i \equiv \mu^i(L(G_i)) = \sum_{j \in \mathcal{I}_Q} \mu(L_{i,j}) = \sum_{j \in \mathcal{I}_Q} \nu(L_{i,j})\chi_j$ is then directly related (but not necessarily equal) to the expected value of the characteristic function. As mentioned earlier, the choice of the characteristic function (Definition 1.2.5) is solely based on the designer's perception of the importance assigned to the individual DFSA states. Therefore, in the setting of the language measure, a supervisor's performance is superior if the supervised plant is more likely to terminate at a *good* marked state and/or less likely to terminate at a *bad* marked state.

1.5 Estimation of Language Measure Parameters

This section presents a recursive algorithm for identification of the language measure parameters (i.e., elements of the event cost matrix $\widetilde{\Pi}$) (see Definition 1.2.6) which, in turn, allows computation of the state transition cost matrix Π (see Definition 1.2.7) and the language measure μ-vector (see Definition 1.2.10). It is assumed that the underlying physical process evolves at two different time scales. In the fast-time scale, i.e., over a short time period, the system is assumed to be an ergodic, discrete Markov process. In the slowly-varying time scale, i.e., over a long period, the system (possibly) behaves as a non-stationary stochastic process. For such a slowly-varying non-stationary process, it might be necessary to redesign the supervisory control policy in real time. In that case, the $\widetilde{\Pi}$-matrix parameters should be updated at selected slow-time epochs.

1.5.1 A Recursive Parameter Estimation Scheme

Let p_{ij} be the transition probability of the event σ_j at the state q_i, i.e.,

$$p_{ij} = \begin{cases} P[\sigma_j|q_i], & \text{if } \exists q \in Q, \ s.t. \ q = \delta(q_i, \sigma_j) \\ 0, & \text{otherwise} \end{cases} \tag{1.35}$$

and its estimate be denoted by the parameter $\hat{p}_{ij}$ that is to be identified from the ensemble of simulation and/or experimental data.

Let a strictly increasing sequence of time epochs of consecutive event occurrence be denoted as:

$$\mathcal{T} \equiv \{t_k : k \in \mathbf{N}_0\} \tag{1.36}$$

where $\mathbf{N}_0$ is the set of non-negative integers. Let the indicator $\psi : \mathbf{N}_0 \times \mathcal{I}_Q \times \mathcal{I}_\Sigma \to \{0, 1\}$ represent the incident of occurrence of an event. For example, if the DFSA was in state q_i at time epoch t_{k-1}, then

$$\psi_{ij}(k) = \begin{cases} 1, & \text{if } \sigma_j \text{ occurs at the time epoch } t_k \in \mathcal{T} \\ 0, & \text{otherwise} \end{cases} \tag{1.37}$$

Consequently, the number of occurrences of any event in the alphabet Σ is represented by $\Psi : \mathbf{N}_0 \times \mathcal{I}_Q \to \{0, 1\}$. For example, if the DFSA was in state q_i at the time epoch t_{k-1}, then

$$\Psi_i(k) = \sum_{j \in \mathcal{I}_\Sigma} \psi_{ij}(k) \tag{1.38}$$

Let $n : \mathbf{N}_0 \times \mathcal{I}_Q \times \mathcal{I}_\Sigma \to \mathbf{N}_0$ represent the cumulative number of occurrences of an event at a state up to a given time epoch. That is, $n_{ij}(k)$ denotes the number of occurrences of the event σ_j at the state q_i up to the time epoch $t_k \in \mathcal{T}$. Similarly, let $N : \mathbf{N}_0 \times \mathcal{I}_Q \to \mathbf{N}_0$ represent the cumulative number of occurrences of any event in the alphabet Σ at a state up to a given time epoch. Consequently,

$$N_i(k) = \sum_{j \in \mathcal{I}_\Sigma} n_{ij}(k) \tag{1.39}$$

A frequency estimator, $\hat{p}_{ij}(k)$, for probability $p_{ij}(k)$ of the event σ_j occurring at the state q_i at the time epoch t_k, is obtained as:

$$\hat{p}_{ij}(k) = \frac{n_{ij}(k)}{N_i(k)}$$
$$\lim_{k \to \infty} \hat{p}_{ij}(k) = p_{ij} \tag{1.40}$$

Convergence of the above limit is justified because the occurrence of an event at a given state of a stationary Markov chain can be treated as an independent and identically distributed random variable.

A recursive algorithm of learning p_{ij} is formulated as a stochastic approximation scheme, starting at the time epoch t_0 with the initial conditions: $\hat{p}_{ij}(0) = 0$ and $n_{ij}(0) = 0$ for all $i \in \mathcal{I}_Q, j \in \mathcal{I}_\Sigma$; and $\Psi_i(0) = 0$ for all $i \in \mathcal{I}_Q$. Starting at $k = 0$, the recursive algorithm runs for $\{t_k : k \geq 1\}$. For example, upon occurrence of an event σ_j at a state q_i, the algorithm is recursively incremented as:

$$n_{ij}(k) = n_{ij}(k-1) + \psi_{ij}(k)$$
$$N_i(k) = N_i(k-1) + \Psi_i(k) \tag{1.41}$$

Next it is demonstrated how the estimates of the language parameters (i.e., the elements of event cost matrix $\widetilde{\Pi}$) are determined from the probability estimates. As stated earlier in Section 1.4, the set of unmodelled events at state q_i, denoted by $\Sigma_i^u \; \forall i \in \mathcal{I}_Q$, accounts for the row-sum inequality: $\sum_j \tilde{\pi}_{ij} < 1$ (see Definition 1.2.6). Then, $\mathrm{P}[\Sigma_i^u] = \theta_i \in (0,1)$ and $\sum_i \tilde{\pi}_{ij} = 1 - \theta_i$. An estimate of the $(i,j)^{th}$ element of the $\widetilde{\Pi}$-matrix, denoted by $\hat{\tilde{\pi}}_{ij}$, is approximated as:

$$\hat{\tilde{\pi}}_{ij}(k) = \hat{p}_{ij}(k)(1-\theta_i) \;\; \forall j \in \mathcal{I}_\Sigma \tag{1.42}$$

Additional experiments on a more detailed automaton model would be necessary to identify the parameters $\theta_i \; \forall i \in \mathcal{I}_Q$. Given that $\theta_i \ll 1$, the problem of conducting additional experimentation can be circumvented by the following approximation:

A single parameter $\theta \approx \theta_i \; \forall i \in \mathcal{I}_Q$, $i \in \mathcal{I}_Q$, such that $0 < \theta \ll 1$, could be selected for convenience of implementation. From the numerical perspective, this option is meaningful because it sets an upper bound on the language measure based on the fact that the sup-norm $\|\mu\|_\infty \le \theta^{-1}$. Note that each row sum in the $\widetilde{\Pi}$-matrix being strictly less than 1, i.e., $\sum_j \tilde{\pi}_{ij} < 1$, is a sufficient condition for finiteness of the language measure.

Theoretically, $\tilde{\pi}_{ij}$ is the asymptotic value of the estimated probabilities $\hat{\tilde{\pi}}_{ij}(k)$ as if the event σ_j occurs infinitely many times at the state q_i. However, dealing with finite amount of data, the objective is to obtain a *good* estimate $\hat{p}_{ij}$ of p_{ij} from independent Bernoulli trials of generating events. Critical issues in dealing with finite amount of data are: (i) how much data are needed; and (ii) when to stop if adequate data are available. The next section 1.5.2 addresses these issues.

1.5.2 Stopping Rules for Recursive Learning

A stopping rule is necessary to find a lower bound on the number of experiments to be conducted for identification of the $\widetilde{\Pi}$-matrix parameters. This section presents two stopping rules that are discussed below.

The first stopping rule is based on an inference approximation having a specified absolute error bound ε with a probability λ. The objective is to achieve a trade-off between the number of experimental observations and the estimation accuracy. A robust stopping rule, similar to [2], is presented below.

A bound on the required number of samples is estimated using the Gaussian structure for binomial distribution that is an approximation of the sum of a large number of independent and identically distributed (i.i.d.) Bernoulli trials of $\hat{\tilde{\pi}}_{ij}(t)$. The central limit theorem yields $\hat{\tilde{\pi}}_{ij} \sim \mathcal{N}(\tilde{\pi}_{ij}, \frac{\tilde{\pi}_{ij}(1-\tilde{\pi}_{ij})}{N})$, where $\mathcal{N}$ indicates normal (or Gaussian) distribution with $E[\hat{\tilde{\pi}}_{ij}] \approx \tilde{\pi}_{ij}$ and

$\text{Var}[\hat{\tilde{\pi}}_{ij}] \equiv \sigma^2 \approx \frac{\tilde{\pi}_{ij}(1-\tilde{\pi}_{ij})}{N}$, provided that the number of samples N is sufficiently large. Let $\Delta = \hat{\tilde{\pi}}_{ij} - \tilde{\pi}_{ij}$, then $\frac{\Delta}{\sigma} \sim \mathcal{N}(0,1)$. Given $0 < \varepsilon \ll 1$ and $0 < \lambda \ll 1$, the problem is to find a bound N_b on the number N of experiments such that $P\{|\Delta| \geq \varepsilon\} \leq \lambda$. Equivalently,

$$P\left\{\frac{|\Delta|}{\sigma} \geq \frac{\varepsilon}{\sigma}\right\} \leq \lambda \tag{1.43}$$

that yields a bound N_b on N as:

$$N_b \geq \left(\frac{\xi^{-1}(\lambda)}{\varepsilon}\right)^2 \tilde{\pi}_{ij}(1-\tilde{\pi}_{ij}) \tag{1.44}$$

where $\xi(x) \equiv 1 - \sqrt{\frac{2}{\pi}} \int_0^x e^{-\frac{t^2}{2}} dt$. Since the parameter $\tilde{\pi}_{ij}$ is unknown, one may use the fact that $\tilde{\pi}_{ij}(1-\tilde{\pi}_{ij}) \leq 0.25$ for every $\tilde{\pi}_{ij} \in [0,1]$ to (conservatively) obtain a bound on N only in terms of the specified parameters ε and λ as:

$$N_b \geq \left(\frac{\xi^{-1}(\lambda)}{2\varepsilon}\right)^2 \tag{1.45}$$

The above estimate of the bound on the required number of samples is less conservative than that obtained from the Chernoff bound and is significantly less conservative than that obtained from Chebyshev bound [13] that does not require the assumption of any specific distribution of Δ except for finiteness of the rth $(r = 2)$ moment.

The second stopping rule, which is an alternative to the first stopping rule, is based on the properties of irreducible stochastic matrices. Following Equation (1.40) and the state transition function δ of the DFSA, the state transition matrix is constructed at the kth iteration as $\mathbf{P}(k)$ that is an irreducible $n \times n$ stochastic matrix under stationary conditions. Similarly, the state probability vector $\mathbf{p}(k) \equiv [p_1(k)\ p_2(k)\ \cdots\ p_n(k)]$ is obtained by following Equation (1.40):

$$p_i(k) = \frac{N_i(k)}{\sum_{j \in \mathcal{I}_Q} N_j(k)} \tag{1.46}$$

The stopping rule makes use of the Perron-Frobenius Theorem to establish a relation between the vector $\mathbf{p}(k)$ and the irreducible stochastic matrix $\mathbf{P}(k)$.

Theorem 1.5.1 *Perron-Frobenius Theorem [17] Let $\mathbf{P}(k)$ be an $n \times n$ irreducible matrix, then there exits an eigenvalue r such that*

1 $r \in \mathbf{R}$ and $r > 0$.

2 r can be associated strictly positive left and right eigenvectors.

3 $r \geq \lambda \; \forall$ eigenvalue $\lambda \neq r$.

4   The eigenvectors associated with r are unique to constant multiples.

5   If $0 \leq B \leq P(k)$ and β is an eigenvalue of B, then $|\beta| \leq |r|$. Moreover, $|\beta| = r$ implies $B = P(k)$.

6 r is a simple root of the characteristic equation of $P(k)$.

Corollary 1.5.1 *Corollary to Perron-Frobenius Theorem*

$$\min_i \sum_{j=1}^n P_{ij}(k) \leq r \leq \max_i \sum_{j=1}^n P_{ij}(k)$$

with equality on either side implying equality throughout.

Since $\mathbf{P}(k)$ is a stochastic matrix, i.e., $\sum_{j=1}^n P_{ij}(k) = 1$, and $\mathbf{P}(k)$ is irreducible, there is a unique eigenvalue $r = 1$ and the left eigenvector corresponding to this eigenvalue (normalized to unity in the sense of absolute sum) represents the state probability vector, provided that the matrix parameters have converged after sufficiently large number of iterations. That is,

$$\| \mathbf{p}(k)\,(\mathbf{I} \; - \; \mathbf{P}(k)) \,\|_\infty \leq \frac{1}{k} \; \rightarrow 0 \; \text{ as } k \rightarrow \infty$$

Equivalently,

$$\| \,(\mathbf{p}(k) \; - \; \mathbf{p}(k+1)) \,\|_\infty \leq \frac{1}{k} \; \rightarrow 0 \; \text{ as } k \rightarrow \infty \qquad (1.47)$$

Taking the expected value of $\| \mathbf{p}(k) \|_\infty$ to be $\frac{1}{n}$, a threshold of $\frac{\eta}{n}$ is specified, where n is the number of states and $0 < \eta \ll 1$ is a constant. A lower bound on the required number of samples is determined from Equation (1.47) as:

$$N_{stop} \equiv Integer \left(\frac{n}{\eta} \right) \qquad (1.48)$$

based on the number of states, n, and the specified tolerance η.

1.6 Usage of the Language Measure

The two methods of language measure computation, presented in Section 1.3, have the same computational complexity, $\mathcal{O}(n^3)$, where n is the number of states of the DFSA. However, each of these two methods offer distinct relative advantages in specific contexts. For example, the recursive solution in Section

1.3.2 might prove very useful for construction of executable codes in real time applications, while the closed form solution in Section 1.3.1 is more amenable for analysis and synthesis of decision and control algorithms. The following two subsections present usage of the language measure for construction of metric spaces of formal languages and synthesis of optimal discrete-event supervisors.

1.6.1 Vector Space of Formal Languages

The language measure can be used to construct a vector space of sublanguages for a given DFSA $G_i \equiv \langle Q, \Sigma, \delta, q_i, Q_m \rangle$. The total variation measure $|\mu|$ [16] induces a metric on this space, which quantifies the distance function between any two sublanguages of $L(G_i)$.

Proposition 1.6.1 *Let $L(G_i)$ be the language of a DFSA $G_i = \langle Q, \Sigma, \delta, q_i, Q_m \rangle$. Let the binary operation of exclusive-OR $\oplus : 2^{L(G_i)} \times 2^{L(G_i)} \to 2^{L(G_i)}$ be defined as:*

$$(K_1 \oplus K_2) = (K_1 \cup K_2) - (K_1 \cap K_2) \tag{1.49}$$

$\forall K_1, K_2 \subseteq L(G_i)$. *Then $(2^{L(G_i)}, \oplus)$ is a vector space over Galois field $GF(2)$.*

Proof. It follows from the properties of exclusive-OR that the algebra $(2^{L(G_i)}, \oplus)$ is an Abelian group where $\emptyset$ is the zero element of the group and the unique inverse of every element $K \subseteq 2^{L(G_i)}$ is K itself because $K_1 \oplus K_2 = \emptyset$ if and only if $K_1 = K_2$. The associative and distributive properties of the vector space follows by defining the scalar multiplication of vectors as: $0 \otimes K = \emptyset$ and $1 \otimes K = K$.

The collection of singleton languages made from each element of $L(G_i)$ forms a basis set of vector space $(2^{L(G_i)}, \oplus)$ over $GF(2)$. It is shown below, how total variation [16] of the signed measure μ can be used to define a metric on above vector space.

Proposition 1.6.2 *Total variation measure $|\mu|$ on $2^{L(G_i)}$ is given by $|\mu|(L) = \sum_{s \in L} |\mu(\{s\})| \ \forall L \subseteq L(G_i)$.*

Proof. The proof follows from the fact that $\Sigma_k |\mu(L_k)|$ attains its supremum for the finest partition of L which consists of the individual strings in L as elements of the partition.

Corollary 1.6.1 (to Theorem 1.6.2) *Let $L(G_i)$ be the language of a DFSA $G_i \equiv \langle Q, \Sigma, \delta, q_i, Q_m \rangle$. For any $K \in 2^{L(G_i)}$, $|\mu|(K) \leq \theta^{-1}$ where $\theta = 1 - \|\Pi\|_\infty$ and Π is the state transition cost matrix of the DFSA.*

Proof. The proof follows from Proposition 1.6.2 and Corollary 1.3.1.

Definition 1.6.1 *Let $L(G_i)$ be the language of a DFSA $G_i \equiv \langle Q, \Sigma, \delta, q_i, Q_m \rangle$. The distance function $d : 2^{L(G_i)} \times 2^{L(G_i)} \rightarrow [0, \infty)$ is defined in terms of the total variation measure as:*

$$d(K_1, K_2) = |\mu|((K_1 \cup K_2) - (K_1 \cap K_2)) \tag{1.50}$$

$\forall K_1, K_2 \subseteq L(G_i)$.

The above distance function $d(\cdot, \cdot)$ quantifies the difference between two supervisors relative to the controlled performance of the DFSA plant.

Proposition 1.6.3 *The distance function defined above is a pseudo-metric on the space $2^{L(G_i)}$*

Proof. Since the total variation of a signed real measure is bounded [16], $\forall K_1, K_2 \subseteq L(G_i), d(K_1, K_2) = |\mu|(K_1 \oplus K_2) \in [0, \infty)$; also by Definition 1.6.1, $d(K_1, K_2) = d(K_2, K_1)$. The remaining property of the triangular inequality follows from the inequality $|\mu|(K_1 \oplus K_2) \leq |\mu|(K_1) + |\mu|(K_2)$ which is based on the fact that $(K_1 \oplus K_2) \subseteq (K_1 \cup K_2)$ and $|\mu|(K_1) \leq |\mu|(K_2) \ \forall K_1 \subset K_2$.

The pseudo-metric $|\mu| : 2^{L(G_i)} \rightarrow [0, \infty)$ can be converted to a metric of the space $(2^{L(G_i)}, \oplus)$ by clustering all languages that have zero total variation measure as the null equivalence class $\mathcal{N} \equiv \{K \in 2^{L(G_i)} : |\mu|(K) = 0\}$. This procedure is conceptually similar to what is done for defining norms in the L_p spaces. In that case, $\mathcal{N}$ contains all sublanguages of $L(G_i)$, which terminate on non-marked states starting from the initial state, i.e., $\mathcal{N} = \{\emptyset \cup (\cup_{q_j \notin Q_m} L_{i,j})\}$. In the sequel, $|\mu|(\cdot)$ is referred to as a metric of the space $2^{L(G_i)}$. Thus, the metric $|\mu|(\cdot)$ can be generated from $d(\cdot, \cdot)$ as: $|\mu|(K) = d(K, J) \ \forall K \in 2^{L(G_i)}$ $\forall J \in \mathcal{N}$. Unlike the norms on vector spaces defined over infinite fields, the metric $|\mu|(\cdot)$ for the vector space $(2^{L(G_i)}, \oplus)$ over $GF(2)$ is not a functional. This interpretation of language as a vector and associating a metric to quantify distance between languages, may be useful for analysis and synthesis of discrete-event supervisory control systems.

1.6.2 Optimal Control of Regular Languages

The (signed) language measure μ could serve as the performance index for synthesis of an optimal control policy (e.g., Sengupta and Lafortune [18]) that maximizes the performance of a controlled sublanguage. The salient concept is succinctly presented below.

Let $\mathcal{S} \equiv \{S^0, S^1, \cdots, S^N\}$ be a set of supervisory control policies for the unsupervised plant automaton G where S^0 is the null controller (i.e., no event is disabled) implying that $L(S^0/G) = L(G)$. Therefore, the controller cost matrix $\Pi(S^0) = \Pi^0$ that is the Π-matrix of the unsupervised plant automaton G. For a supervisor $S^k, k \in \{1, 2, \cdots, N\}$, the control policy is required to selectively disable certain controllable events so that the following (elementwise) inequality holds: $\Pi^k \equiv \Pi(S^k) \leq \Pi^0$ and $L(S^k/G) \subseteq L(G), \forall S^k \in \mathcal{S}$.

The task is to synthesize an optimal cost matrix $\mathbf{\Pi}^* \leq \mathbf{\Pi}^0$ that maximizes the performance vector $\mu^* \equiv [\mathbf{I} - \mathbf{\Pi}^*]^{-1}\mathbf{X}$, i.e., $\mu^* \geq \mu^k \equiv [\mathbf{I} - \mathbf{\Pi}^k]^{-1}\mathbf{X}$ $\forall$ $\mathbf{\Pi}^k \leq \mathbf{\Pi}^0$ where the inequalities are implied elementwise. Research work in this direction has been reported in recent literature [5] [4] and is also described in detail in Chapters 2 and 3.

1.7 An Application Example

Fu et al. [5] have adopted the closed form method of language measure as a performance index for optimal supervisory control of a twin-engine unmanned aircraft that is used for surveillance and data collection. The language measure has been computed and verified based on both the techniques given in Section 1.3. Engine health and operating conditions, which are monitored in real time based on observed data, are classified into three mutually exclusive and exhaustive categories: *good*; *unhealthy* (but operable); and *inoperable*. In the event of any observed abnormality, the supervisor may decide to continue or abort the mission. The finite state automaton model of the plant in Figure 1.2 has 13 states (excluding the dump state), of which three are marked states, and nine events, of which four are controllable and the remaining five are uncontrollable. All events are assumed to be observable. The states and events of the plant model are listed in Table 1.1 and Table 1.2, respectively. The state transition function δ and the state-based event cost $\tilde{\pi}_{ij}$ (see Definition 1.2.6) are entered simultaneously in Table 1.3. The values of $\tilde{\pi}_{ij}$ were selected by extensive experiments on engine simulation models and were also based on experience of gas turbine engine operation and maintenance. The dump state and any transitions to the dumped state are not shown in Table 1.3. The elements of the characteristic vector (see Definition 1.2.5) were chosen as signed real weights based on the perception of each marked state's role on the engine performance.

The χ values of the 13 states in Table 7.1 are given by the characteristic vector $\mathbf{X} = [0\ 0\ 0\ 0\ 0\ 0\ 0\ 0\ 0\ 0\ -0.05\ +0.25\ -1.0]^T$. These parameters are selected by the designer based on his/her perception of each marked state's role in the system performance. As the states 1 to 10 are not marked, the first 10 elements of the characteristic vector $\mathbf{X}$ are zeros. The implication is that event strings terminating at states 1 to 10 have no bearing on the system performance and hence have zero measure. The state 12 is a *good* marked state having a positive χ value and the *bad* marked states 11 and 13 have negative χ values. Therefore, event strings terminating at state 12 have positive measure and those terminating at states 11 and 13 have negative measure.

Three supervisory controllers were designed independently using a graphical interactive package [21] based on the following specifications:

1. Specification #1: At least one of the two engines must be in *good* condition for mission continuation.

Table 1.1. Plant Automaton States

State	Description
1	Safe in base
2	Mission executing - two good engines
3	One engine unhealthy during mission execution
4	Mission executing - one good and one unhealthy engine
5	Both engines unhealthy during mission execution
6	One engine good and one engine inoperable
7	Mission execution with two unhealthy engines
8	Mission execution with only one good engine
9	One engine unhealthy and one engine inoperable
10	Mission execution with only one unhealthy engine
11	Mission aborted /not completed (Bad Marked State)
12	Mission successful (Good Marked State)
13	Aircraft destroyed (Bad Marked State)

Table 1.2. Plant Event Alphabet

Event	Event Description	Controllable Event
s	start and take-off	√
b	a good engine becoming unhealthy	
t	an unhealthy engine becoming inoperable	
v	a good engine becoming inoperable	
k	keep engine(s) running	√
a	mission abortion	√
f	mission completion	
d	destroyed aircraft	
l	landing	√

2. Specification #2: None of the two engines must be in *inoperable* condition for mission continuation.
3. Specification#3: Both engines must be in *good* condition for mission continuation.
4. Optimal Control: The control policy is optimized with with zero disabling cost, as described in Chapter 2.

The supervised plant automata under specifications #1, #2, #3, and optimal control are displayed in Figures 1.3, 1.4, 1.5, and 1.6, respectively, where dashed lines indicate disabled controllable events. Notice that none of the controllable events are disabled in the unsupervised plant (see Figure 1.2) and the four supervisors disable different sets of controllable events, as seen in Figures 1.3 to 1.6.

The performance measure μ_1 (i.e., with the initial state 1) of the unsupervised (i.e., no disabling of control events) plant is 0.0823 and for three super-

Table 1.3. State Transition and Event Cost Matrix

	s	b	t	v	k	a	f	d	l
1	0.50 (2)					0.02 (1)			
2		0.05 (3)		0.01) (6)			0.80 (12)	0.10 (13)	
3					0.45 (4)	0.45 (11)			
4		0.12 (5)	0.16 (6)	0.10 (9)			0.50 (12)	0.12 (13)	
5					0.45 (7)	0.45 (11)			
6					0.45 (8)	0.45 (11)			
7			0.25 (9)				0.50 (12)	0.20 (13)	
8		0.20 (9)		0.01 (13)			0.3 (12)	0.4 (13)	
9					0.45 (10)	0.45 (11)			
10			0.35 (13)				0.20 (12)	0.40 (13)	
11									0.95 (1)
12									0.95 (1)
13									

vised plants under specifications #1, #2, #3, and the optimally supervised plant are evaluated to be: 0.0807, 0.0822, 0.0840, and 0.0850, respectively. Therefore, the performance of the supervised plant under specifications #1, #2, and #3 is inferior, similar, and superior, respectively, to that of the unsupervised plant. As expected, the optimal supervisor has better performance than that of Supervisor #3. Notice that Supervisor #3 does disable the controllable event k from the state 3 to state 4 and the optimal supervisor does not. That is, the optimal supervisor allows continuing operation of an unhealthy engine while the remaining engine is in good condition.

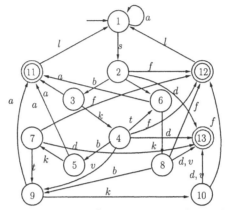

Fig. 1.2. Unsupervised Plant (i.e., no disabling of controllable events)

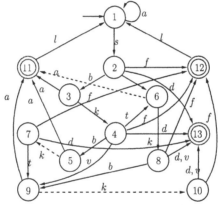

Fig. 1.3. Supervised Plant under Specification #1

1.8 Summary, Conclusions, and Future Research

This chapter reviews the concept, formulation and validation of a signed real measure for a regular language and its sublanguages based on the principles of measure theory and automata theory. While the domain of measure μ, i.e., $2^{L(G_i)}$, is partially ordered, its range, which is a subset of $\mathbf{R} \equiv (-\infty, \infty)$, becomes totally ordered. As a result, the relative performance of different supervisors can be quantitatively evaluated in terms of the real signed measure of the controlled sublanguages. Positive weights are assigned to *good* marked states and negative weights to *bad* marked states so that a controllable supervisor is rewarded (penalized) for deleting strings terminating at *bad* (*good*) marked states. In order to evaluate and compare the performance of different

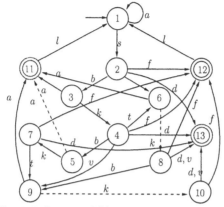

Fig. 1.4. Supervised Plant under Specification #2

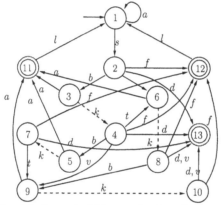

Fig. 1.5. Supervised Plant under Specification #3

supervisors a common quantitative tool is required. To this effect, the proposed procedure computes the measure of the controlled sublanguage generated by a supervisor using the event cost and characteristic function assigned for the unsupervised plant. Cost assignment to each event based on the state, where it is generated, has been shown similar to the conditional probability of the event. On the other hand, the characteristic function is chosen based on the designer's perception of the individual state's impact on the system performance. Two techniques are presented to compute the language measure for a DFSA. One of them yields a closed form solution that is obtained as the unique solution of a set of linearly independent algebraic equations. The other is based on a recursive procedure. The computational complexity of both lan-

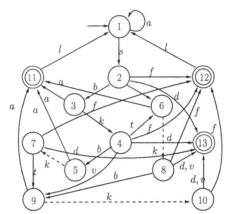

Fig. 1.6. Optimally Supervised Plant

guage measure algorithms is of polynomial order in the number of states of the DFSA.

1.8.1 Recommendations for Future Research

Further research is recommended for development of language measures if some of the events are delayed or may not be observable at all. This research is necessary for synthesis of supervisory algorithms under delayed or partial observation of events. Another critical issue is how to extend the language measure for timed automaton, especially if the events are observed with varying delays at different states. Another research topic that may also be worth investigating is: how to extend the field $GF(2)$, over which the vector space of languages has been defined, to richer fields like the set of real numbers.

It would be challenging to extend the concept of (regular) language measure for languages higher up in the Chomsky Hierarchy [9] such as context free and context sensitive languages. This extension would lead to controller synthesis when the plant dynamics is modelled by non-regular languages such as the Petri-Net.

Other areas of future research include applications of the language measure in anomaly detection, model identification, model order reduction, and analysis and synthesis of interfaces between the continuously-varying and discrete-event spaces in the language-measure setting.

Appendix A
Measure Theory

This appendix introduces the concepts of standard measure-theoretic quantities that are used to establish the language measure in the main body of this chapter.

Definition A.1 *A σ-algebra M of a nonempty language $L(G_i) \subseteq \Sigma^*$ is a collection of subsets of $L(G_i)$ which satisfies the following three conditions:*

(i) $L(G_i) \in M$;
(ii) If $K \in M$, then $(L(G_i) - K) \in M$;
(iii) $\cup_{j=1}^{\infty} K_j \in M$ if $K_{kj} \in M \ \forall j$.

Definition A.2 *An at most countable collection $\{L_k\}$ of members of a $\sigma-$ algebra M is a partition of a member $L \in M$ if $L = \cup_k L_k$ and $L_k \cap L_j = \emptyset$ $\forall k \neq j$.*

Definition A.3 *Let M be a σ-algebra of $L(G_i)$. Then, the set function $\mu :$ $M \to \mathbf{R} \equiv (-\infty, +\infty)$, is called a signed real measure if the following two conditions are satisfied [16]:*

(i) $\mu(\emptyset) = 0$;
(ii) $\mu(\cup_{j=1}^{\infty} L_j) = \sum_{k=1}^{\infty} \mu(L_j)$ *for every partition $\{L_k\}$ on any member $L \in M$.*

Note that, unlike a positive measure (e.g., the Lebesgue measure), μ is finite such that the series in part (ii) of Definition A.3 converges absolutely in $\mathbf{R}$ and the result is independent of any permutation of the terms under union.

Definition A.4 *Relative to the signed real measure μ, a sublanguage $L \in M$ is defined to be:*

(i) null, denoted as $L = 0$, if $\mu(L \cap J) = 0, \forall J \in M$;
(ii) positive, denoted as $L > 0$, if $L \neq 0$ and $\mu(L \cap J) \geq 0, \forall J \in M$;
(iii) negative. denoted as $L < 0$, if $L \neq 0$ and $\mu(L \cap J) \leq 0, \forall J \in M$.

Definition A.5 *Total variation $|\mu|$ on a σ-algebra M is defined as:*

$$|\mu|(L) = \sup \sum_k |\mu(L_k)| \tag{1.51}$$

$\forall L \in M$ *where the supremum is taken over all partitions $\{L_k\}$ of L.*

Proposition A.1 *Total variation measure $|\mu|$ of any regular language L is nonnegative and finite i.e., $|\mu|(L) \in [0, \infty)$.*

The proof follows from standard theorems on complex measures [16].

Total variation can be, in general, defined for complex measures [16] but it is restricted to a signed real measure in this chapter. The total variation of a real signed measure μ, can be represented as $|\mu| = \mu^+ + \mu^-$ where μ^+ and μ^- are called the positive and negative variation of μ and are defined as

$$\mu^+ = \frac{1}{2}(|\mu| + \mu) \qquad \text{and} \qquad \mu^- = \frac{1}{2}(|\mu| - \mu) \tag{1.52}$$

Both μ^+ and μ^- are positive measures on M. It also follows from above equation that $\mu = \mu^+ - \mu^-$. This representation of μ as the difference of positive measure μ^+ and μ^- is known as the Jordan Decomposition of μ [6].

Proposition A.2 *Every sublanguage $L \in M$ can be partitioned as: $L = L^0 \cup L^+ \cup L^-$ where the mutually exclusive sublanguages L^0, L^+ and L^- are called null, positive, and negative, respectively, relative to a signed real measure μ .*

The proof is based on the Hahn Decomposition Theorem [6]. As a consequence of above result, the following relations hold $\forall L \in M$ for positive and negative variations:

$$\mu^+(L) = \mu(L \cap L^+) \qquad \text{and} \qquad \mu^-(L) = -\mu(L \cap L^-) \tag{1.53}$$

Appendix B
Convergence of the Language Measure μ

This appendix establishes necessary and sufficient conditions for finiteness of the measure μ, based on certain properties of non-negative matrices which are stated without proof. The reader is referred to [12] and [17] for details of these results.

Definition B.1 *Let A and B be real square matrices of the same order n. Then, the following are notations:*

$$\begin{array}{ll} A \geq B & if\ a_{ij} \geq b_{ij}, \forall i, j \\ A > B & if\ A \geq B,\ A \neq B \\ A \gg B & if\ a_{ij} > b_{ij}, \forall i, j \end{array}$$

If the matrix A satisfies the condition $A > 0$, i.e. the null matrix, then A is called a nonnegative matrix and if the condition $A \gg 0$ is satisfied, then it is called positive matrix.

Definition B.2 *A square matrix A of order n is cogradient to a matrix E if for some permutation matrix P, $PAP^T = E$. A is reducible if it is cogradient to:*

$$E = \begin{bmatrix} B & 0 \\ C & D \end{bmatrix}$$

where B and C are square matrices, or if $n = 1$ and $A = 0$. Otherwise, A is irreducible.

It follows from above definition that a positive matrix is always irreducible.

Proposition B.1 *A nonnegative matrix A is irreducible if and only if for every (i, j) there exists a natural number q such that*

$$a_{ij}^{(q)} > 0$$

where $a_{ij}^{(q)}$ denote the (i, j) element of A^q.

Proposition B.2 *If $A \geq 0$ is irreducible and $B \geq 0$, then $A + B$ is irreducible.*

Another characterization of irreducibility of a nonnegative square matrix has a graph-theoretic interpretation. This relationship can help to determine under what conditions the given finite state automaton G, which represents the controlled or uncontrolled plant model is irreducible by looking at the connectivity of its states. The following definitions are needed to To arrive at this conclusion.

Definition B.3 *The associated directed graph , $G(A)$ of a square matrix A of order n, consists of n vertices $P_1, P_2 \ldots, P_n$ where an edge leads from P_i to P_j if and only if $a_{ij} \neq 0$.*

Definition B.4 *A directed graph G is strongly connected if for any ordered pair (P_i, P_j) of vertices of G, there exists a sequence of edges which leads from P_i to P_j.*

Proposition B.3 *Given a matrix A, it is irreducible if and only if $G(A)$ is strongly connected.*

If A is a nonnegative square matrix, then the following is a result regarding relationship between the spectral radius (i.e., maximum absolute eigenvalue) ρ of nonnegative matrices.

Proposition B.4 *If $0 \leq A \leq B$ and $A + B$ is irreducible then $\rho(A) < \rho(B)$.*

Definition B.5 *A square matrix S of order n is called (row) stochastic if it satisfies*

$$s_{ij} \geq 0, \qquad \sum_{j=1}^{n} s_{ij} = 1 \qquad 1 \leq i \leq n \tag{1.54}$$

Proposition B.5 *The maximum eigenvalue of a stochastic matrix S is one, i.e $\rho(S) = 1$. A nonnegative matrix A is stochastic if and only if e is an eigenvector of A corresponding to eigenvalue one, where e is the vector all of whose entries are equal to one.*

In order to show that $(I - \Pi)^{-1}$ is invertible it suffices to show that $\rho(\Pi) < 1$.

Theorem B.1 *If $\rho(\Pi) < 1$ then at least for one i, $1 \leq i \leq n$, $\sum_{j=1}^{n} \pi_{ij} < 1$.*

Proof. Proof follows from the fact that if $\sum_{j=1}^{n} \pi_{ij} = 1$ $\forall i$, then Π would be a stochastic matrix by Definition B.5. Hence by proposition B.5 $\rho(\Pi) = 1 \Rightarrow$ $(I - \Pi)^{-1}$ is not invertible.

Theorem B.2 *If $\sum_{j=1}^{n} \pi_{ij} < 1, \forall i$ s.t $1 \leq i \leq n$, then $\rho(\Pi) < 1$.*

Proof. Let $\theta_i = (1 - \sum_{j=1}^{n} \pi_{ij})/n > 0$. Let S be a matrix of order n which is defined in the following manner: $s_{ij} = \theta_i + \pi_{ij}$, $1 \leq i, j \leq n$. It is clear that $S >> 0$ and hence S is irreducible. Also S is a stochastic matrix and by proposition B.5, $\rho(S) = 1$. Since $0 \leq \Pi < S$ and $\Pi + S$ is irreducible by proposition B.2, it follows that $\rho(\Pi) < \rho(S) = 1$ from proposition B.4.

The above sufficiency condition is more strict than the necessary condition required in Theorem B.1. However, the necessary condition is not sufficient as seen from the following example:

$$\Pi = \begin{pmatrix} 0.2 & 0 & 0.8 & 0 \\ 0 & 0.2 & 0.3 & 0.5 \\ 0.5 & 0 & 0.5 & 0 \\ 0.1 & 0.2 & 0.4 & 0 \end{pmatrix}$$

This matrix Π satisfies conditions as required in Theorem B.1, but $\rho(\Pi) = 1$. It is possible to relax the strict inequality $\sum_{j=1}^{n} \pi_{ij} < 1$, $\forall i$ in Theorem B.2, but with additional conditions on structure of Π matrix. For example, under such relaxed conditions, if $\Pi + S$ is irreducible, then still $\rho(\Pi) < 1$. This follows from the fact that application of Proposition B.4, just requires irreducibility of $\Pi + S$. In order to determine the irreducibility of a matrix, the graph-theoretic interpretation, described earlier, can be a useful tool.

References

1. C.G. Cassandras and S. Lafortune, *Introducrion to discrete event systems*, Kluwer Academic, 1999.
2. A. Doucet, S. Godsill, and C. Andrieu, *On sequential monte carlo sampling methods for bayesian filtering*, Statistics and Computing **10** (2000), no. 3, 197–208.
3. V. Drobot, *Formal languages and automata theory*, Computer Science Press, 1989.
4. J. Fu, A. Ray, and C.M. Lagoa, *Optimal control of regular languages with event disabling cost*, Proceedings of American Control Conference, Denver, Colorado, June 2003, pp. 1691–1695.

5. J. Fu, A. Ray, and C.M. Lagoa, *Unconstrained optimal control of regular languages*, Automatica **40** (2004), no. 4, 639–648.
6. P.R. Halmos, *Measure theory, 2nd ed.*, Springer-Verlag, 1974.
7. J. E. Hopcroft, R. Motwani, and J. D. Ullman, *Introduction to automata theory, languages, and computation, 2nd ed.*, Addison-Wesley, 2001.
8. R. Kumar and V. Garg, *Modeling and control of logical discrete event systems*, Kluwer Academic, 1995.
9. J. C. Martin, *Introduction to languages and the theory of computation, 2nd ed.*, McGraw-Hill, 1997.
10. A.W. Naylor and G.R. Sell, *Linear operator theory in engineering and science*, Springer-Verlag, 1982.
11. S. Phoha, E. Peluso, and R.L. Culver, *A high fidelity ocean sampling mobile network (samon) simulator*, IEEE Journal of Oceanic Engineering, Special Issue on Autonomous Ocean Sampling Networks **26** (2002), no. 4, 646–653.
12. R. J. Plemmons and A. Berman, *Nonnegative matrices in the mathematical science*, Academic Press, New York, 1979.
13. M. Pradhan and P. Dagum, *Optimal monte carlo estimation of belief network inference*, Twelfth Conference on Uncertainty in Artificial Intelligence (Portland, OR), 1996, pp. 446–453.
14. P.J. Ramadge and W.M. Wonham, *Supervisory control of a class of discrete event processes*, SIAM J. Control and Optimization **25** (1987), no. 1, 206–230.
15. A. Ray and S. Phoha, *Signed real measure of regular languages for discrete-event automata*, Int. J. Control **76** (2003), no. 18, 1800–1808.
16. W. Rudin, *Real and complex analysis, 3rd ed.*, McGraw-Hill, New York, 1987.
17. E. Senata, *Non-negative matrices*, John Wiley, New York, 1973.
18. R. Sengupta and S. Lafortune, *An optimal control theory for discrete event systems*, SIAM J. Control and Optimization **36** (1998), no. 2, 488–541.
19. A. Surana and A. Ray, *Signed real measure of regular languages*, Demonstratio Mathematica **37** (2004), no. 2, 485–503.
20. X. Wang and A. Ray, *A language measure for performance evaluation of discrete-event supervisory control systems*, Applied Mathematical Modelling **28** (2004), no. 9, 817–833.
21. X. Wang, A. Ray, S. Phoha, and J. Liu, *J-des: A graphical interactive package for analysis and synthesis of discrete event systems*, Proceedings of American Control Conference (Denver, Colorado), June 2003, pp. 3405–3410.

2

Optimal Supervisory Control of Regular Languages

Asok Ray[1], Jinbo Fu[2], and Constantino Lagoa[3]

[1] The Pennsylvania State University, University Park, PA 16802 axr2@psu.edu
[2] The Pennsylvania State University, University Park, PA 16802 jxf293@psu.edu
[3] The Pennsylvania State University, University Park, PA 16802
lagoa@engr.psu.edu

Summary. This chapter presents optimal supervisory control of dynamical systems that can be represented by deterministic finite state automaton (DFSA) models. The performance index for the optimal policy is obtained by combining a measure of the supervised plant language with (possible) penalty on disabling of controllable events. The signed real measure quantifies the behavior of controlled sublanguages based on a state transition cost matrix and a characteristic vector as reported in Chapter 1 and earlier publications. Synthesis of the optimal control policy requires at most n iterations, where n is the number of states of the DFSA model generated from the unsupervised plant language. The computational complexity of the optimal control synthesis is polynomial in n. Syntheses of the control algorithms are illustrated with two application examples.

Key words: Discrete Event Supervisory Control, Optimal Control, Language Measure

2.1 Introduction

In a seminal paper [8], Ramadge and Wonham pioneered the concept of discrete-event supervisory control of finite-state automata (equivalently, regular languages). A plant, supervised under a control policy, is represented by a sublanguage of the unsupervised plant language, which could be different under different supervisors if they are constrained to satisfy dissimilar specifications. Such a set of supervised plant sublanguages are, in general, partially ordered; it is necessary to establish a quantitative measure for total ordering of their respective performance. To address this issue, Wang and Ray [12], and their colleagues [9] [11] have developed a signed measure of regular languages. Details of the language measure have been presented in Chapter 1.

Optimal control of regular languages has been proposed by several researchers based on different assumptions. Some of these researchers have attempted to quantify the controller performance using different types of cost assigned to the individual events. Passino and Antsaklis [7] proposed path costs associated with state transitions and hence optimal control of a discrete event system is equivalent to following the shortest path on the graph representing the uncontrolled system. Kumar and Garg [5] made use of the concept of payoff and control costs that are incurred only once regardless of the number of times the system visits the state associated with the cost. Consequently, the resulting cost is not a function of the dynamic behavior of the plant. Brave and Heymann [1] introduced the concept of optimal attractors in discrete-event control. Sengupta and Lafortune [10] used control cost in addition to the path cost in optimization of the performance index for trade-off between finding the shortest path and reducing the control cost. Although costs were assigned to the events, no distinction was made for events generated at (or leading to) different states that could be "good" or "bad".

These optimal control strategies have addressed performance enhancement of discrete-event control systems without a quantitative measure of languages.

Recently, Fu et al. [4] have proposed a state-based method for optimal control of regular languages by selectively disabling controllable events so that the resulting optimal policy can be realized as a controllable supervisor. The performance index of the optimal policy is a signed real measure of the supervised sublanguage, which is expressed in terms of a cost matrix and a characteristic vector [12] [9] [11], but it does not assign any additional penalty for event disabling. In a follow-up publication, Fu et al. [3] extended their earlier work on optimal control to include the cost of (controllable) event disabling. The rationale is that, without the event disabling cost, an optimal supervisor makes the best trade-off between reaching good states and avoiding bad states, and achieves optimal performance in terms of the language measure of the supervised plant. However, another supervisor that has a slightly inferior performance relative to the above optimal controller may only require disabling of fewer or some other controllable events, which is much less difficult to achieve. Therefore, with due consideration to event disabling, the second controller might be preferable.

The work, reported in this chapter, augments and consolidates the theory and applications of optimal supervisory control of regular languages, which have been reported in an informal structure in previous publications [4] [3]. The performance index for the optimal control policy proposed in this chapter is obtained by combining a real signed measure of the supervised plant language with the cost of disabled event(s). Starting with the (regular) language of an unsupervised plant automaton, the optimal control policy makes a trade-off between the measure of the supervised sublanguage and the associated event disabling cost to achieve the best performance. Like any other optimization procedure, it is possible to choose different performance indices to arrive at different optimal policies for discrete event supervisory control. Nevertheless, usage of the language measure provides a systematic procedure for precise comparative evaluation of different supervisors so that the optimal control policy(ies) can be conclusively identified.

The theoretical results are presented with formal proofs and are supported by two application examples. The first application is on supervisory control of a twin-engine aircraft for its safe and reliable operation. The second application is on decision and control of a multiprocessor decoding system for efficient operation. The major contribution of this chapter is conceptualization, formulation and illustration of a quantitative method for analysis and synthesis of optimal supervisory control policies for plant dynamics that can be captured by regular languages. From the above perspectives, the performance index for the optimal control policy, presented in this chapter, is obtained by combining the measure of the supervised plant language with the cost of disabled event(s).

This chapter is organized in six sections including the present one. Section 2.2 briefly reviews the previous work on language measure. Section 2.3 presents the optimal control policy without the event disabling cost. Section 2.4 modifies the performance index to include the event disabling cost and formulates the algorithm of the optimal control policy with event disabling cost as an extension of Section 2.3. Section 2.5 presents two application examples to illustrate the concepts of optimal control without and with event disabling cost. The chapter is summarized and concluded in Section 2.6 along with recommendations for future work.

2.2 Brief Review of Language Measure

This section briefly reviews the concept of signed real measure of regular languages [12] [9] [11], which is described in details in Chapter 1. Let the plant behavior be modelled as a deterministic finite state automaton (DFSA):

$$G_i \equiv (Q, \Sigma, \delta, q_i, Q_m) \tag{2.1}$$

where Q is the finite set of states with $|Q| = n$, and $q_i \in Q$ is the initial state; Σ is the (finite) alphabet of events with $|\Sigma| = m$; the Kleene closure of Σ is denoted as Σ^* that is the set of all finite-length strings of events including the empty string ε; the (possibly partial) function $\delta : Q \times \Sigma \rightarrow Q$ represents state transitions and $\hat{\delta} : Q \times \Sigma^* \rightarrow Q$ is an extension of δ; and $Q_m \subseteq Q$ is the set of marked (i.e., accepted) states.

Definition 2.2.1 *The language $L(G_i)$ generated by a DFSA G initialized at the state $q_i \in Q$ is defined as:*

$$L(G_i) = \{s \in \Sigma^* \mid \hat{\delta}(q_i, s) \in Q\} \tag{2.2}$$

Definition 2.2.2 *The language $L_m(G_i)$ marked by a DFSA G_i initialized at the state $q_i \in Q$ is defined as:*

$$L_m(G_i) = \{s \in \Sigma^* \mid \hat{\delta}(q_i, s) \in Q_m\} \tag{2.3}$$

The language $L(G_i)$ is partitioned as the non-marked and the marked languages, $L^o(G_i) \equiv L(G_i) - L_m(G_i)$ and $L_m(G_i)$, consisting of event strings that, starting from $q_i \in Q$, terminate at one of the non-marked states in $Q - Q_m$ and one of the marked states in Q_m, respectively. The set Q_m is further partitioned into Q_m^+ and Q_m^-, where Q_m^+ contains all *good* marked states that are desired to be terminated on and Q_m^- contains all *bad* marked states that one may not want to terminate on, although it may not always be possible to avoid the bad states while attempting to reach the good states. The marked language $L_m(G_i)$ is further partitioned into $L_m^+(G_i)$ and $L_m^-(G_i)$

consisting of good and bad strings that, starting from q_i, terminate on Q_m^+ and Q_m^-, respectively.

A signed real measure $\mu : 2^{L(G_i)} \to \mathbf{R} \equiv (-\infty, \infty)$ is constructed for quantitative evaluation of every event string $s \in L(G_i)$. The language $L(G_i)$ is decomposed into null, i.e., $L^o(G_i)$, positive, i.e., $L_m^+(G_i)$, and negative, i.e., $L_m^-(G_i)$ sublanguages.

Definition 2.2.3 *The language of all strings that, starting at a state $q_i \in Q$, terminates on a state $q_j \in Q$, is denoted as $L(q_i, q_j)$. That is,*

$$L(q_i, q_j) \equiv \{s \in L(G_i) : \hat{\delta}(q_i, s) = q_j\}. \tag{2.4}$$

Definition 2.2.4 *The characteristic function that assigns a signed real weight to state-partitioned sublanguages $L(q_i, q_j), i = 1, 2, \ldots, n, j = 1, 2, \ldots, n$ is defined as: $\chi : Q \to [-1, 1]$ such that*

$$\chi(q_j) \in \begin{cases} [-1, 0) \ if \ q_j \in Q_m^- \\ \{0\} \quad if \ q_j \notin Q_m \\ (0, 1] \ if \ q_j \in Q_m^+ \end{cases} \tag{2.5}$$

Definition 2.2.5 *The event cost is conditioned on a DFSA state at which the event is generated, and is defined as $\tilde{\pi} : \Sigma^* \times Q \to [0, 1]$ such that $\forall q_j \in Q, \forall \sigma_k \in \Sigma, \forall s \in \Sigma^*$,*

(1) $\tilde{\pi}[\sigma_k, q_j] \equiv \tilde{\pi}_{jk} \in [0, 1); \ \sum_k \tilde{\pi}_{jk} < 1;$
(2) $\tilde{\pi}[\sigma, q_j] = 0$ if $\delta(q_j, \sigma)$ is undefined; $\tilde{\pi}[\epsilon, q_j] = 1;$
(3) $\tilde{\pi}[\sigma_k s, q_j] = \tilde{\pi}[\sigma_k, q_j] \ \tilde{\pi}[s, \delta(q_j, \sigma_k)].$

The event cost matrix is defined as:

$$\widetilde{\Pi} = \begin{bmatrix} \tilde{\pi}_{11} & \tilde{\pi}_{12} & \cdots & \tilde{\pi}_{1m} \\ \tilde{\pi}_{21} & \tilde{\pi}_{22} & \cdots & \tilde{\pi}_{2m} \\ \vdots & \vdots & \ddots & \vdots \\ \tilde{\pi}_{n1} & \tilde{\pi}_{n2} & \cdots & \tilde{\pi}_{nm} \end{bmatrix} \tag{2.6}$$

and is referred to as the $\widetilde{\Pi}$-matrix in the sequel.

An application of the induction principle to part (3) in Definition 2.2.5 shows $\tilde{\pi}[st, q_j] = \tilde{\pi}[s, q_j]\tilde{\pi}[t, \hat{\delta}(q_j, s)]$. The condition $\sum_k \tilde{\pi}_{jk} < 1$ provides a sufficient condition for the existence of the real signed measure as discussed in [11] along with additional comments on the physical interpretation of the event cost.

The next task is to formulate a measure of sublanguages of the plant language $L(G_i)$ in terms of the signed characteristic function χ and the non-negative event cost $\tilde{\pi}$.

Definition 2.2.6 *The signed real measure μ of a singleton string set $\{s\} \subseteq L(q_i, q_j) \subseteq L(G_i) \in 2^{\Sigma^*}$ is defined as:*

$$\mu(\{s\}) \equiv \tilde{\pi}(s, q_i)\chi(q_j) \quad \forall s \in L(q_i, q_j). \tag{2.7}$$

The signed real measure of $L(q_i, q_j)$ is defined as:

$$\mu(L(q_i, q_j)) \equiv \sum_{s \in L(q_i, q_j)} \mu(\{s\}) \tag{2.8}$$

and the signed real measure of a DFSA G_i, initialized at the state $q_i \in Q$, is denoted as:

$$\mu_i \equiv \mu(L(G_i)) = \sum_j \mu(L(q_i, q_j)) \tag{2.9}$$

Definition 2.2.7 *The state transition cost, $\pi : Q \times Q \to [0, 1)$, of the DFSA G_i is defined as follows:*

$$\forall q_i, q_j \in Q, \pi_{ij} = \begin{cases} \sum_{\sigma \in \Sigma} \tilde{\pi}[\sigma, q_i], & \text{if } \delta(q_i, \sigma) = q_j \\ 0 & \text{if } \{\delta(q_i, \sigma) = q_j\} = \emptyset. \end{cases} \tag{2.10}$$

Consequently, the $n \times n$ state transition cost Π-matrix is defined as:

$$\Pi = \begin{bmatrix} \pi_{11} & \pi_{12} & \cdots & \pi_{1n} \\ \pi_{21} & \pi_{22} & \cdots & \pi_{2n} \\ \vdots & \vdots & \ddots & \vdots \\ \pi_{n1} & \pi_{n2} & \cdots & \pi_{nn} \end{bmatrix} \tag{2.11}$$

Wang and Ray [12], and Surana and Ray [11] have shown that the measure $\mu_i \equiv \mu(L(G_i))$ of the language $L(G_i)$, with the initial state q_i, can be expressed as: $\mu_i = \sum_j \pi_{ij} \mu_j + \chi_i$ where $\chi_i \equiv \chi(q_i)$. Equivalently, in vector notation: $\bar{\mu} = \Pi\bar{\mu} + \bar{\chi}$ where the measure vector $\bar{\mu} \equiv [\mu_1 \ \mu_2 \ \cdots \ \mu_n]^T$ and the characteristic vector $\bar{\chi} \equiv [\chi_1 \ \chi_2 \ \cdots \ \chi_n]^T$. From the perspective of constructing an optimal control policy, salient properties of the state transition cost matrix Π are delineated below.

Property 1: *Following Definition 2.2.5, there exists $\theta \in (0, 1)$ such that the induced infinity norm $\|\Pi\|_\infty \equiv \max_i \sum_j \pi_{ij} = 1 - \theta$. The matrix operator $[I - \Pi]$ is invertible implying that the inverse $[I - \Pi]^{-1}$ is a bounded linear operator [6] with its induced infinity norm $\|[I - \Pi]^{-1}\|_\infty \leq \theta^{-1}$. Therefore, the language measure vector can be expressed as: $\bar{\mu} = [I - \Pi]^{-1}\bar{\chi}$, where $\bar{\mu} \in \mathbb{R}^n$, and computational complexity [11] of the measure is $O(n^3)$.*

Property 2: *The matrix operator $[I - \Pi]^{-1} \geq 0$ elementwise. By Taylor series expansion, $[I - \Pi]^{-1} = \sum_{k=0}^{\infty} [\Pi]^k$ and $[\Pi]^k \geq 0$ because $\Pi \geq 0$.*

Property 3: *The determinant $Det[I - \Pi]$ is real positive because the eigenvalues of the real matrix $[I - \Pi]$ appear as real or complex conjugates* *and they have positive real parts. Hence, the product of all eigenvalues of* $[I - \Pi]$ *is real positive.*

Property 4: *An affine operator* $T : \mathbf{R}^n \rightarrow \mathbf{R}^n$ *can be defined as:* $T \bar{\nu} = \Pi \bar{\nu} + \bar{\chi}$ *for any arbitrary* $\nu \in \mathbf{R}^n$. *As* Π *is a contraction,* T *is also a contraction. Since* $\mathbf{R}^n$ *is a Banach space, there exists a unique fixed point [6] of* T, *i.e., the measure vector* $\bar{\mu}$ *satisfying the condition* $T \bar{\mu} = \bar{\mu}$. *Therefore, the language measure vector* $\bar{\mu}$ *is uniquely determined as:* $\bar{\mu} = [I - \Pi]^{-1} \bar{\chi}$, *which can be interpreted as the unique fixed point of a contraction operator.*

2.3 Optimal Control without Event Disabling Cost

This section presents the theoretical foundations of the optimal supervisory control of deterministic finite state automata (DFSA) plants by selectively disabling controllable events so that the resulting optimal policy can be realized as a controllable supervisor [4]. The plant model is first modified to satisfy the specified operational constraints, if any; this model is referred to as the *unsupervised or open loop* plant in the sequel. Then, starting with the (regular) language of the unsupervised plant, the optimal policy maximizes the performance of the controlled sublanguage of the supervised plant without any further constraints. The performance index of the optimal policy is a signed real measure of the supervised sublanguage, described in Section 2.2, which is expressed in terms of a state transition cost matrix Π and a characteristic vector $\bar{\chi}$, but it does not assign any additional penalty for event disabling.

Let $\mathcal{S} \equiv \{S^0, S^1, \cdots, S^N\}$ be the finite set of all supervisory control policies that selectively disable controllable events of the unsupervised plant DFSA G and can be realized as regular languages. Denoting $\Pi^k \equiv \Pi(S^k)$, $k \in \{1, 2, \cdots, N\}$, the supervisor S^0 is the null controller (i.e., no event is disabled) implying that $L(S^0/G) = L(G)$. The controller cost matrix $\Pi(S^0) = \Pi^0 \equiv \Pi^{plant}$ that is the Π-matrix of the unsupervised plant automaton G. For a supervisor S^i, $i \in \{1, 2, \cdots, N\}$, the control policy selectively disables certain controllable events at specific state(s); consequently, the corresponding elements of the $\tilde{\Pi}$-matrix (see Definition 2.2.4) become zero. Therefore, the inequalities, $\pi_{ij} \geq 0$ and $\sum_j \pi_{ij} < 1$, hold and $L(S^k/G) \subseteq L(G)$ $\forall S^k \in \mathcal{S}$. The language measure vector of a supervised plant $L(S^k/G)$ is expressed as:

$$\bar{\mu}^k \equiv [I - \Pi^k]^{-1} \bar{\chi} \tag{2.12}$$

where the jth element of the vector $\bar{\mu}^k$ is denoted as μ_j^k. In the sequel, $\bar{\mu}^*$ is chosen to be the performance measure for the optimal control policy without event disabling cost.

Proposition 2.3.1 *Let j be such that $\mu_j^k = \min\limits_{\ell \in \{1,2,\cdots,n\}} \mu_\ell^k$. If $\mu_j^k \leq 0$, then $\chi_j \leq 0$; and if $\mu_j^k < 0$, then $\chi_j < 0$.*

Proof. The DFSA satisfies the identity $\mu_j^k = \sum\limits_{\ell \in \{1,2,\cdots,n\}} \pi_{j\ell}^k \mu_\ell^k + \chi_j$ that leads to the inequality $\mu_j^k \geq (\sum_\ell \pi_{j\ell}^k)\mu_j^k + \chi_j \Rightarrow (1 - \sum_\ell \pi_{j\ell}^k)\mu_j^k \geq \chi_j$. The proof follows from $(1 - \sum_\ell \pi_{j\ell}^k) > 0$ (see Definitions 2.2.5 and 2.2.7).

Corollary 2.3.1 *Let $\mu_j^k = \max\limits_{\ell \in \{1,2,\cdots,n\}} \mu_\ell^k$. If $\mu_j^k \geq 0$, then $\chi_j \geq 0$ and if $\mu_j^k > 0$, then $\chi_j > 0$.*

Proof. The proof is similar to that of Proposition 2.3.1.

Proposition 2.3.2 *Given $\Pi(S^k) = \Pi^k$ and $\mu^k \equiv [I - \Pi^k]^{-1}\bar{\chi}$, let Π^{k+1} be generated from Π^k for $k \geq 0$ by disabling or re-enabling the appropriate controllable events as follows: $\forall i,j \in \{1, 2, \cdots, n\}$, ij^{th} element of Π^{k+1} is modified as:*

$$\pi_{ij}^{k+1} \begin{cases} \geq \pi_{ij}^k & if \ \mu_j^k > 0 \\ = \pi_{ij}^k & if \ \mu_j^k = 0 \\ \leq \pi_{ij}^k & if \ \mu_j^k < 0 \end{cases} \tag{2.13}$$

and $\Pi^k \leq \Pi^0 \ \forall k$. Then, $\bar{\mu}^{k+1} \geq \bar{\mu}^k$ elementwise and equality holds if and only if $\Pi^{k+1} = \Pi^k$.

Proof. It follows from the the properties of the measure vector $\bar{\mu}$ that:

$$\begin{aligned} \bar{\mu}^{k+1} - \bar{\mu}^k &= \left([I - \Pi^{k+1}]^{-1} - [I - \Pi^k]^{-1}\right)\bar{\chi} \\ &= [I - \Pi^{k+1}]^{-1}\left([I - \Pi^k] - [I - \Pi^{k+1}]\right)[I - \Pi^k]^{-1}\bar{\chi} \\ &= [I - \Pi^{k+1}]^{-1}\left(\Pi^{k+1} - \Pi^k\right)\bar{\mu}^k \end{aligned}$$

Defining the matrix $\Delta^k \equiv \Pi^{k+1} - \Pi^k$, let the jth column of Δ^k be denoted as Δ_j^k. Then, $\Delta_j^k \leq 0$ if $\mu_j^k < 0$ and $\Delta_j^k \geq 0$ if $\mu_j^k \geq 0$, and the remaining columns of Δ^k are zero vectors. This implies that: $\Delta^k \bar{\mu}^k = \sum_j \Delta_j^k \mu_j^k \geq 0$. Since $\Pi^k \leq \Pi^0 \ \forall k$, $[I - \Pi^{k+1}]^{-1} \geq 0$ elementwise. Then, it follows that $[I - \Pi^{k+1}]^{-1}\Delta^k\bar{\mu}^k \geq 0 \Rightarrow \bar{\mu}^{k+1} \geq \bar{\mu}^k$. For $\mu_j^k \neq 0$ and Δ^k as defined above, $\Delta^k\bar{\mu}^k = 0$ if and only if $\Delta^k = 0$. Then, $\Pi^{k+1} = \Pi^k$ and $\bar{\mu}^{k+1} = \bar{\mu}^k$.

Corollary 2.3.2 *For a given state q_j, let $\mu_j^k < 0$ and Π^{k+1} be generated from Π^k by disabling controllable events that lead to the state q_j. Then, $\mu_j^{k+1} < 0$.*

Proof. Since only j^{th} column of $[I - \Pi^{k+1}]$ is different from that of $[I - \Pi^k]$ and the remaining columns are the same, the jth row of the cofactor matrix of $[I - \Pi^{k+1}]$ is the same as that of the cofactor matrix of $[I - \Pi^k]$. Therefore,

$$Det\,[I - \Pi^{k+1}]\mu_j^{k+1} = Det\,[I - \Pi^k]\mu_j^k$$

Since both determinants are real positive by Property 3 of the Π-matrix in Section 2.2, μ_j^k and μ_j^{k+1} have the same sign.

In Proposition 2.3.2, some elements of the jth column of Π^k are decreased (or increased) by disabling (or re-enabling) controllable events that lead to the states q_j for which $\mu_j^k < 0$ (or $\mu_j^k \geq 0$). Next it is shown that an optimal supervisor can be achieved to yield best performance in terms of the language measure.

Proposition 2.3.3 *Iterations of event disabling and re-enabling lead to a cost matrix Π^* that is optimal in the sense of maximizing the performance vector $\bar{\mu}^* \equiv [I - \Pi^*]^{-1}\bar{\chi}$ elementwise.*

Proof. Let us consider an arbitrary cost matrix $\Pi^\dagger \leq \Pi^0$ and $\bar{\mu}^\dagger \equiv [I - \Pi^\dagger]^{-1}\bar{\chi}$. It will be shown that $\bar{\mu}^\dagger \leq \bar{\mu}^*$. Let us rearrange the elements of the $\bar{\mu}^*$-vector such that $\bar{\mu}^* = [\underbrace{\mu_1^* \cdots \mu_\ell^*}_{\geq 0} \mid \underbrace{\mu_{\ell+1}^* \cdots \mu_n^*}_{< 0}]^T$ and the cost matrices $\Pi^\dagger$ and Π^* are also rearranged in the order in which the $\bar{\mu}^*$-vector is arranged.

According to Proposition 2.3.2, no controllable event leading to states q_k, $k = 1, 2, \cdots, \ell$, is disabled and all controllable events leading to states q_k, $k = \ell + 1, \ell + 2, \cdots, n$, are disabled. Therefore, the elements in the first ℓ columns of Π^* are the same as those of the Π^0 and only the elements in the last $(n-\ell)$ columns are decreased to the maximum permissible extent by disabling all controllable events. In contrast, the columns of $\Pi^\dagger$ are reduced by an arbitrary choice. Therefore, defining $\Delta\Pi^\dagger \equiv [\Pi^\dagger - \Pi^*]$, the first ℓ columns of $\Delta\Pi^\dagger \leq 0$ and the last $(n - \ell)$ columns of $\Delta\Pi^\dagger \geq 0$.

Since $\bar{\mu}^* = [\underbrace{\mu_1^* \cdots \mu_\ell^*}_{\geq 0} \mid \underbrace{\mu_{\ell+1}^* \cdots \mu_n^*}_{< 0}]^T$ and $[I - \Pi^\dagger]^{-1} \geq 0$ elementwise, and $\bar{\mu}^\dagger - \bar{\mu}^* = [I - \Pi^\dagger]^{-1} [\Pi^\dagger - \Pi^*] \mu^*$, it follows that

$$\bar{\mu}^\dagger - \bar{\mu}^* = \underbrace{[I - \Pi^\dagger]^{-1}}_{\geq 0} \left(\underbrace{\sum_{j=1}^{\ell} Col_j \cdot \mu_j^*}_{\leq 0} + \underbrace{\sum_{j=\ell+1}^{n} Col_j \cdot \mu_j^*}_{\leq 0} \right) \leq 0$$

where Col_j indicates j^{th} column of the matrix $[\Pi^\dagger - \Pi^*]$.
Therefore, $\bar{\mu}^\dagger \leq \bar{\mu}^*$ for any arbitrary choice of $0 \leq \Pi^\dagger \leq \Pi^0$.

Proposition 2.3.4 *The control policy induced by the optimal Π^*-matrix in Proposition 2.3.3 is unique in the sense that the controlled language is most permissive (i.e., least restrictive) among all controller(s) having the best performance.*

Proof. Disabling controllable event(s) leading to a state q_j with performance measure $\mu_j^* = 0$ does not alter the performance vector $\bar{\mu}^*$. The optimal control does not disable any controllable event leading to the states with zero performance. This implies that, among all controllers with the identical performance $\bar{\mu}^*$, the control policy induced by the Π^*-matrix is most permissive.

Propositions 2.3.3 and 2.3.4 suffice to conclude that the Π^*-matrix yields the most permissive controller with the best performance $\bar{\mu}^*$. The optimal control policy (without event disabling cost) can be realized as:

- All controllable events leading to the states q_j, for which $\mu_j^* < 0$, are disabled;
- All controllable events leading to the states q_j, for which $\mu_j^* \geq 0$, are enabled.

2.3.1 Optimal Policy Construction without Event Disabling Cost

A procedure is proposed for construction of the optimal control policy that maximizes the performance of the controlled language of DFSA (without event disabling cost), starting from any initial state $q_i \in Q$. Let G_i be a DFSA plant model without any constraint (i.e., operational specifications) and have the state transition cost matrix of the unsupervised plant as: $\Pi^0 \equiv \Pi^{plant} \in \mathbf{R}^{n \times n}$ and the characteristic vector as: $\bar{\chi} \in \mathbf{R}^n$. Then, the performance vector at $k = 0$ is given as: $\bar{\mu}^0 = [\mu_1^0 \ \mu_2^0 \ \cdots \ \mu_n^0]^T = (I - \Pi^0)^{-1} \bar{\chi}$, where the jth element μ_j^0 of the vector $\bar{\mu}^0$ is the performance of the language, with state q_j as the initial state. Then, $\mu_j^0 < 0$ implies that, if the state q_j is reached, then the plant will yield bad performance thereafter. Intuitively, the control system should attempt to prevent the automaton from reaching q_j by disabling all controllable events that lead to this state. Therefore, the optimal control algorithm starts with disabling all controllable events that lead to every state q_j for which $\mu_j^0 < 0$. This is equivalent to reducing all elements of the corresponding columns of the Π^0-matrix by disabling those controllable events. In the next iteration, i.e., $k = 1$, the updated cost matrix Π^1 is obtained as: $\Pi^1 = \Pi^0 - \Delta^0$ where $\Delta^0 \geq 0$ (the inequality being implied elementwise) is composed of event costs corresponding to all controllable events that have been disabled. Using Proposition 2.3.2, $\bar{\mu}^0 \leq \bar{\mu}^1 \equiv [I - \Pi^1]^{-1} \bar{\chi}$. Although all controllable events leading to every state corresponding to a negative element of μ^1 are disabled, some of the controllable events that were disabled at $k = 0$ may now lead to states corresponding to positive elements of $\bar{\mu}^1$. Performance could be further enhanced by re-enabling these controllable events. For $k \geq 1$, $\Pi^{k+1} = \Pi^k + \Delta^k$ where $\Delta^k \geq 0$ is composed of the state transition costs of all re-enabled controllable events at k.

If $\bar{\mu}^0 \geq 0$, i.e., there is no state q_j such that $\mu_j^0 < 0$, then the plant performance cannot be improved by event disabling and the null controller S^0

(i.e., no disabled event) is the optimal controller for the given plant. Therefore, the cases are considered where $\mu_j^0 < 0$ for some state q_j.

Starting with $k = 0$ and $\Pi^0 \equiv \Pi^{plant}$, the control policy is constructed by the following two-step procedure:

Step 1: For every state q_j for which $\mu_j^0 < 0$, disable controllable events leading to q_j. Now, $\Pi^1 = \Pi^0 - \Delta^0$, where $\Delta^0 \geq 0$ is composed of event costs corresponding to all controllable events, leading to q_j for which $\mu_j^0 < 0$, which have been disabled at $k = 0$.

Step 2: For $k \geq 1$, if $\mu_j^k \geq 0$, re-enable all controllable events leading to q_j, which were disabled in Step 1. The cost matrix is updated as: $\Pi^{k+1} = \Pi^k + \Delta^k$ for $k \geq 1$, where $\Delta^k \geq 0$ is composed of event costs corresponding to all currently re-enabled controllable events. The iteration is terminated if no controllable event leading to q_j remains disabled for which $\mu_j^k > 0$. At this stage, the optimal performance $\bar{\mu}^* \equiv [I - \Pi^*]^{-1} \bar{\chi}$.

Proposition 2.3.5 *The number of iterations needed to arrive at the optimal control law without event disabling cost does not exceed the number, n, of states of the DFSA.*

Proof. Following Proposition 2.3.2, the sequence of performance vectors $\{\Pi^k\}$ in successive iterations of the two-step procedure is monotonically increasing. The first iteration at $k = 0$ disables controllable events following Step 1 of the two-step procedure in Section 2.3.1. During each subsequent iteration in Step 2, the controllable events leading to at least one state are re-enabled. When Step 2 is terminated, there remains at least one negative element, $\mu_j^k < 0$ by Corollary 2.3.2. Therefore, as the number of iterations in Step 2 is at most $n - 1$, the total number of iterations to complete the two-step procedure does not exceed n.

Since each iteration in the synthesis of the optimal control requires a single Gaussian elimination of n unknowns from n linear algebraic equations, computational complexity of the control algorithm is polynomial in n.

2.4 Optimal Control with Event Disabling Cost

This section presents the optimal control policy with (state-based) event disabling cost by including the cost of all (controllable) events, disabled by the supervisor, in the performance cost; the disabling cost is incurred each time the event is disabled at the state. As the cost of disabled event(s) approaches zero, the optimal control policy with event disabling cost converges to the optimal control policy without event disabling cost, described in Section 2.3.

Definition 2.4.1 *Let the cost of disabling a (controllable) event σ_j that causes transition from q_i be denoted as c_{ij} where $c_{ij} \in [0, 1]$. The $(n \times m)$ disabling cost matrix is denoted as $C = [c_{ij}]$.*

Since the (controllable) supervisor never disables any uncontrollable event, the entries c_{ij} for uncontrollable events have no importance. For implementation, they can be set to an arbitrarily large positive real number $M < \infty$.

Definition 2.4.2 *The action of disabling a (controllable) event σ_j at state q_i by a supervisor S is defined as:*

$$d_{ij}^S = \begin{cases} 1 & if \ \sigma_j \ is \ disabled \\ 0 & otherwise \end{cases} \tag{2.14}$$

The $(n \times m)$ action matrix of disabling controllable events by a supervisor S is denoted as: $D^S = [d_{ij}^S]$.

Definition 2.4.3 *The event disabling cost characteristic of a supervisor S that selectively disables controllable events σ_j at state q_i is defined as:*

$$\gamma_i^S = \sum_{j: \ d_{ij}^S = 1} c_{ij} \, \tilde{\pi}_{ij} \tag{2.15}$$

The disabling cost characteristic is proportional to event cost of the controllable event disabled by the supervisor S.

The $(n \times 1)$ disabling cost characteristic vector of a supervisor S is denoted as: $\bar{\gamma}^S \equiv [\gamma_1^S \ \gamma_2^S \ \cdots \ \gamma_n^S]^T$.

Definition 2.4.4 *The modified characteristic of a state $q_i \in Q$ is defined as:*

$$\chi_i^S \equiv \chi_i - \gamma_i^S. \tag{2.16}$$

The $(n \times 1)$ modified characteristic vector under a supervisor S is defined as:

$$\bar{\chi}^S \equiv \bar{\chi} - \bar{\gamma}^S \tag{2.17}$$

where $\bar{\chi}^S \equiv [\chi_i^S \ \chi_i^S \ \cdots \ \chi_n^S]^T$.

Definition 2.4.5 *The disabling cost measure vector under a supervisor S is defined as:*

$$\bar{\theta}^S \equiv [I - \Pi^S]^{-1} \bar{\gamma}^S. \tag{2.18}$$

with θ_i^S being the ith element of $\bar{\theta}^S$, which is the disabling cost incurred by the state.

Definition 2.4.6 *The performance measure vector of a supervisor S is defined as:*

$$\bar{\eta}^S \equiv [I - \Pi^S]^{-1} \bar{\chi}^S \qquad (2.19)$$

with η_i^S being the ith element of $\bar{\eta}^S$.

The performance index vector $\bar{\eta}^S$ of a supervisor S can be interpreted as the difference between the measure vector $\bar{\mu}^S$ of the supervised language $L(S/G)$ of the DFSA G and the respective disabling cost measure vector $\bar{\theta}^S$. That is,

$$\bar{\eta}^S = \bar{\mu}^S - \bar{\theta}^S. \qquad (2.20)$$

Following the approach taken for optimal control without event disabling cost in Section 2.3, let $S \equiv \{S^0, S^1, \cdots, S^N\}$ be the finite set of supervisory control policies that can be realized as regular languages. For a supervisor $S^k \in S$, the control policy selectively disables certain controllable events. Consequently, the corresponding elements of the $\widetilde{\Pi}$-matrix become zero and those of the event disabling characteristic vector $\bar{\gamma}^S$ are entered in the modified characteristic vector $\bar{\chi}^S$ as seen in Definition 2.4.4; therefore, $L(S^k/G) \subseteq L(G)\ \forall S^k \in S$. Denoting $\Pi^k \equiv \Pi(S^k)$, $k \in \{1, 2, \cdots, N\}$, the performance measure vector (with event disabling cost) of the supervised plant $L(S^k/G)$ is expressed as:

$$\bar{\eta}^k \equiv [I - \Pi^k]^{-1}(\bar{\chi} - \bar{\gamma}^k) \qquad (2.21)$$

where $\bar{\eta}^k \equiv \bar{\eta}^{s^k}$; $\bar{\gamma}^k \equiv \bar{\gamma}^{s^k}$; and j^{th} element of the vector $\bar{\eta}^k$ is denoted as η_j^k. The null supervisor S^0 (i.e., no disabled event) has zero disabling cost, i.e., $\bar{\gamma}^0 = 0$ and consequently $\bar{\eta}^0 = \bar{\mu}^0$. The construction of optimal policy is extended to include the event disabling cost.

2.4.1 Optimal Policy Construction with Event Disabling Cost

This subsection formulates an optimal control policy with event disabling cost, which maximizes all elements of the performance vector $\bar{\eta}^S$ of the supervised language of a DFSA G with event cost matrix $\widetilde{\Pi} \in \mathbf{R}^{n \times m}$; state transition cost matrix $\Pi \in \mathbf{R}^{n \times n}$; characteristic vector $\bar{\chi} \in \mathbf{R}^n$; and the disabling cost matrix $C \in \mathbf{R}^{n \times m}$. For the unsupervised plant, the initial conditions of the optimal synthesis procedure are set as: $\Pi^0 \equiv \Pi^{plant}$; $\bar{\chi}^0 = \bar{\chi}$; $\bar{\gamma}^0 = 0$; $D^0 = \mathbf{0}$ (no event disabled so far). For optimal control without event disabling cost in Section 2.3.1, all controllable events that lead to states q_ℓ, for which $\mu_\ell^0 < 0$ are first disabled. Subsequently, for $k \geq 1$, all previously disabled controllable events leading to q_j are re-enabled if $\mu_j^k \geq 0$. In contrast, for optimal control with event disabling cost, the first action is to disable all controllable events σ_j leading to states q_ℓ for which $\eta_\ell^0 < -c_{ij}$ with $\delta(q_i, \sigma_j) = q_\ell$. Subsequently, for $k \geq 1$, these disabled events are re-enabled if $\eta_\ell^k \geq -c_{ij}$. The rationale for this procedure is that disabling of controllable events leading to states with small negative performance may not be advantageous because of incurring additional event disabling cost.

The control policy with event disabling cost is constructed by the following two-step procedure:

Step 1: Starting at $k = 0$ and $\Pi^0 \equiv \Pi^{plant}$, disable all controllable events σ_j, leading to each state q_ℓ if the inequality: $\eta_\ell^0 < -c_{ij}$ with $\delta(q_i, \sigma_j) = q_\ell$ is satisfied. The algorithm for dealing with this inequality is delineated below:

• If the inequality is not satisfied for any single case, stop the iterative procedure. No event disabling can improve the plant performance beyond that of the open loop plant, i.e., the null supervisor S^0 achieves optimal control.

• If the inequality is satisfied for at least one case, disable the qualified event(s) and update the state transition cost matrix to $\Pi^1 \leq \Pi^0$ (elementwise); the disabling matrix to D^1 for generating the cost characteristic function $\bar{\gamma}^1$; and the modified characteristic vector $\bar{\chi}^1 \equiv \bar{\chi} - \bar{\gamma}^1$. Go to Step 2.

Step 2: This step starts at $k = 1$ and the performance measure vector for $k \geq 1$ is:

$$\bar{\eta}^k \equiv [I - \Pi^k]^{-1}\bar{\chi}^k = [I - \Pi^k]^{-1}(\bar{\chi} - \bar{\gamma}^k)$$

The algorithm at Step 2 re-enables all previously (i.e., at $k \geq 0$) disabled controllable events σ_j that lead to states q_ℓ if the inequality $\eta_\ell^k \geq -c_{ij}$ with $\delta(q_i, \sigma_j) = q_\ell$ is satisfied. The algorithm for dealing with this inequality is as follows:

• If the inequality is not satisfied for any single case, an optimal control is achieved and the iterative procedure is complete. No further event re-enabling can improve the controlled plant performance beyond that of the current supervisor that is the optimal controller.

• If the inequality is satisfied for at least one case, re-enable all qualified events and update the state transition cost matrix to $\Pi^{k+1} \geq \Pi^k$ (elementwise); the disabling matrix to D^k; the cost characteristic function to $\bar{\gamma}^{k+1}$; and the modified characteristic vector $\bar{\chi}^{k+1} \equiv \bar{\chi} - \bar{\gamma}^{k+1}$. Update $k \leftarrow (k+1)$ and repeat Step 2 until the inequality $\eta_\ell^k \geq -c_{ij}$ with $\delta(q_i, \sigma_j) = q_\ell$ is not satisfied for all j and ℓ. Then, the current supervisor is optimal in terms of the performance measure in Definition 2.4.6.

The above procedure for optimal control with event disabling cost is an extension of that without event disabling cost described in Section 2.3.1. For zero event disabling cost, the two procedures become identical. Following the rationale of Proposition 2.3.5, the computational complexity of the control synthesis with disabling cost is also polynomial in n.

The underlying theory of unconstrained optimal control with event disabling cost is presented as two additional propositions, which simultaneously maximize all elements of the performance vector $\bar{\eta}$.

Proposition 2.4.1 *For all supervisors S^k in the iterative procedure, $\bar{\eta}^{k+1} \geq \bar{\eta}^k$ elementwise.*

Proof. *Given $\bar{\chi}^k \equiv \bar{\chi} - \bar{\gamma}^k$ and $\bar{\eta}^k \equiv [I - \Pi^k]^{-1} \bar{\chi}^k$, let us denote the change in event disabling characteristic vector as:*

$$\bar{\omega}^k \equiv \bar{\gamma}^{k+1} - \bar{\gamma}^k = \bar{\chi}^k - \bar{\chi}^{k+1}.$$

Notice that, elementwise

$$\bar{\omega}^k \begin{cases} > 0 & \text{for event disabling} \\ \leq 0 & \text{for event re} - \text{enabling} \end{cases}$$

The performance increment at iteration k is given by:

$$
\begin{aligned}
\bar{\eta}^{k+1} - \bar{\eta}^k &= \left[I - \Pi^{k+1}\right]^{-1} \bar{\chi}^{k+1} - \left[I - \Pi^k\right]^{-1} \bar{\chi}^k \\
&= \left[I - \Pi^{k+1}\right]^{-1} \left[\bar{\chi}^k - \bar{\omega}^k\right] - \left[I - \Pi^k\right]^{-1} \bar{\chi}^k \\
&= \left(\left[I - \Pi^{k+1}\right]^{-1} - \left[I - \Pi^k\right]^{-1}\right) \bar{\chi}^k - \left[I - \Pi^{k+1}\right]^{-1} \bar{\omega}^k \\
&= \left[I - \Pi^{k+1}\right]^{-1} \left[\Pi^{k+1} - \Pi^k\right] \left[I - \Pi^k\right]^{-1} \bar{\chi}^k - \left[I - \Pi^{k+1}\right]^{-1} \bar{\omega}^k \\
&= -\left\{\left[I - \Pi^{k+1}\right]^{-1} \left[\Pi^k - \Pi^{k+1}\right] \bar{\eta}^k + \left[I - \Pi^{k+1}\right]^{-1} \bar{\omega}^k\right\}
\end{aligned}
$$

At $k = 0$, the state transition cost matrix changes from Π^0 to Π^1 as a result of disabling selected controllable events leading to states with sufficiently negative performance. Let the ith column of a matrix A be denoted as $(A)_i$; ij^{th} element of a matrix A be denoted as $(A)_{ij}$, and the ith element of a vector v be denoted as $(v)_i$; and let ℓ and j satisfy the following conditions: $\delta(q_\ell, \sigma_j) = q_p$ and $d_{\ell j}^{S^k} \neq d_{\ell j}^{S^{k+1}}$.

Then, $\Pi^1 \leq \Pi^0$; $\omega_\ell^0 = \sum_j c_{\ell j} \left\{\dagger \Pi^0 - \dagger \Pi^1\right\}_{\ell j}$; and

$$
\begin{aligned}
(\bar{\eta}^1 - \bar{\eta}^0)_i &= -\left(\left[I - \Pi^1\right]^{-1} \left[\Pi^0 - \Pi^1\right] \bar{\eta}^0 - \left[I - \Pi^1\right]^{-1} \bar{\omega}^0\right)_i \\
&= -\sum_\ell \left(\left[I - \Pi^1\right]^{-1}\right)_{i\ell} \left(\sum_p \left(\sum_j (\tilde{\pi}_{\ell j} \eta_p^0 + c_{\ell j} \tilde{\pi}_{\ell j})\right)\right) \\
&= -\sum_\ell \left(\left[I - \Pi^1\right]^{-1}\right)_{i\ell} \left(\sum_p \left(\sum_j \tilde{\pi}_{\ell j} (\eta_p^0 + c_{\ell j})\right)\right)
\end{aligned}
$$

Since $\left[I - \Pi^1\right]^{-1} \geq 0$ elementwise and event disabling requires $(\eta_p^0 + c_{\ell j}) < 0$ for all admissible ℓ, j and p (see Step 1 of the control policy with event disabling cost in Section 2.4.1), it follows from the above equation that $\bar{\eta}^1 - \bar{\eta}^0 \geq 0$ elementwise.

Next, iterations $k \geq 1$ are considered, for which some of the events disabled at $k = 0$ are (possibly) re-enabled.

$$\omega_\ell^k = -\sum_j c_{\ell j} \left(\tilde{\Pi}^{k+1} - \tilde{\Pi}^k \right)_{\ell j}$$

$$\left(\bar{\eta}^{k+1} - \bar{\eta}^k \right)_i = \left(\left[I - \Pi^{k+1} \right]^{-1} \left[\Pi^{k+1} - \Pi^k \right] \bar{\eta}^k - \left[I - \Pi^{k+1} \right]^{-1} \bar{\omega}^k \right)_i$$

$$= \sum_\ell \left(\left[I - \Pi^{k+1} \right]^{-1} \right)_{i\ell} \left(\sum_p \left(\sum_j \tilde{\pi}_{\ell j} (\eta_p^k + c_{\ell j}) \right) \right)$$

Since $\left[I - \Pi^k \right]^{-1} \geq 0$ *elementwise and event re-enabling requires* $(\eta_p^k + c_{\ell j}) \geq 0$ *for all admissible* ℓ, j *and* p *(see Step 1 of the control policy with event disabling cost in Section 2.4.1), it follows from the above equations that* $\bar{\eta}^{k+1} - \bar{\eta}^k \geq 0$ *for* $k \geq 0$.

Proposition 2.4.2 *The supervisor generated upon completion of the algorithm in Section 2.4.1 is optimal in terms of the performance in Definition 2.4.6.*

Proof. Based on the algorithm in Section 2.4.1, let the supervisor S^* *be synthesized by disabling and re-enabling certain controllable events at selected states. It is to be shown that* S^* *is optimal in the following sense. The performance* $\bar{\eta}$ *of any (controllable) supervisor* $S \in \mathcal{S}$ *is not superior to the performance* $\bar{\eta}^*$ *of* S^*, *i.e.,* $\bar{\eta}^* \geq \bar{\eta} \quad \forall S \in \mathcal{S}$.

Let an arbitrary supervisor $S \in \mathcal{S}$ *disable ceratin controllable events* σ_j *at selected states* q_ℓ, *which are not disabled by* S^*. *Then, by Step 1 of the control policy with event disabling cost in Section 2.4.1, it follows that* $(\eta_p^* + c_{\ell j}) \geq 0$ *with* $\delta(q_\ell, \sigma_j) = q_p$. *Let the same supervisor* S *enable some other controllable events* σ_k *at selected states* q_ℓ, *leading to state* q_r, *which are disabled by* S^*, *i.e.,* $(\eta_r^* + c_{\ell k}) < 0$ *with* $\delta(q_\ell, \sigma_k) = q_r$. *In non-trivial cases,* ℓ, j *and* k *satisfy the conditions:* $d_{\ell j}^{S^*} \neq d_{\ell j}^S$ *and* $d_{\ell k}^{S^*} \neq d_{\ell k}^S$.

Denoting the differences between event disabling characteristic vectors and the state transition cost matrices of S^* *and* S, *respectively, as:*

$$\bar{\omega} \equiv \bar{\gamma}^S - \bar{\gamma}^{S^*} = \bar{\chi}^{S^*} - \bar{\chi}^S$$

$$\Delta \equiv \Pi^{S^*} - \Pi^S$$

the difference between corresponding performance vectors is obtained as:

$$\bar{\eta}^* - \bar{\eta} = \left[I - \Pi^{S^*} \right]^{-1} \bar{\chi}^{S^*} - \left[I - \Pi^S \right]^{-1} \bar{\chi}^S$$

$$= \left[I - \Pi^{S^*} \right]^{-1} \bar{\chi}^{S^*} - \left[I - \Pi^S \right]^{-1} \left[\bar{\chi}^{S^*} - \bar{\omega} \right]$$

$$= \left(\left[I - \Pi^{S^*} \right]^{-1} - \left[I - \Pi^S \right]^{-1} \right) \bar{\chi}^{S^*} + \left[I - \Pi^S \right]^{-1} \bar{\omega}$$

$$= \left[I - \Pi^S \right]^{-1} \left[\Pi^{S^*} - \Pi^S \right] \left[I - \Pi^{S^*} \right]^{-1} \bar{\chi}^{S^*} + \left[I - \Pi^S \right]^{-1} \bar{\omega}$$

$$= \left[I - \Pi^S \right]^{-1} \left[\Pi^{S^*} - \Pi^S \right] \bar{\eta}^{S^*} + \left[I - \Pi^S \right]^{-1} \bar{\omega}$$

$$= \left[I - \Pi^S \right]^{-1} \left(\Delta \, \bar{\eta}^{S^*} + \bar{\omega} \right)$$

Since the matrix $\left[I - \Pi^S\right]^{-1} \geq 0$, *it suffices to show that* $\left(\bar{\eta}^{S^*} + \bar{\omega}\right) \geq 0$
to prove $\left(\bar{\eta}^{S^*} - \bar{\eta}^S\right) \geq 0$ *elementwise. The proof will make use of the fact*
that $\pi_{i\ell} = \sum_u \tilde{\pi}_{iu}$, *where* $\sigma_u \in \Sigma$ *and* $\delta(\sigma_u, q_i) = q_\ell$ *for all pairs of states*
q_i *and* q_ℓ *(see Definitions 2.2.5 and 2.2.7) under any given supervisor S.*
Then, the matrix Δ *can be partitioned as two sets of columns such that, for*
$\ell \in \{1, 2, \cdots, n\}$, *(nonzero) elements in each row of these column sets are*
obtained as:

$$\Delta_{\ell p} = \sum_{j : d_{\ell j}^S = 1} \tilde{\pi}_{\ell j} \text{ and } \Delta_{\ell r} = - \sum_{k : d_{\ell k}^{S^*} = 1} \tilde{\pi}_{\ell k}$$

where the subscript p depends on both ℓ *and* j, *and the subscript r depends*
on both ℓ *and* k. *The product of the matrix* $\Delta \in \Re^{n \times n}$ *and the performance*
vector $\bar{\eta}^* \in \Re^n$ *is derived as:*

$$(\Delta \, \bar{\eta}^*)_\ell \equiv \sum_p \Delta_{\ell p} \eta_p^* + \sum_r \Delta_{\ell r} \eta_r^*$$

$$= \sum_p \left(\sum_j \tilde{\pi}_{\ell j} \, \eta_p^* \right) - \sum_r \left(\sum_k \tilde{\pi}_{\ell k} \eta_r^* \right)$$

Similarly, the change in the event disabling characteristic vector is expressed
as:

$$(\bar{\omega})_\ell = \sum_i \tilde{\pi}_{\ell i} c_{\ell i} \left(d_{\ell i}^S - d_{\ell i}^{S^*} \right)$$

$$= \sum_p \left(\sum_j \tilde{\pi}_{\ell j} c_{\ell j} \right) - \sum_r \left(\sum_k \tilde{\pi}_{\ell k} c_{\ell k} \right)$$

Summing the above two expressions yields the ℓth *element of the vector*
$(\Delta \bar{\eta}^{S^*} + \bar{\omega})$ *as:*

$$(\Delta \bar{\eta}^* + \bar{\omega})_\ell$$

$$= \sum_p \left(\sum_j \tilde{\pi}_{\ell j} \, \eta_p^* \right) - \sum_r \left(\sum_k \tilde{\pi}_{\ell k} \, \eta_r^{S^*} \right) + \sum_p \left(\sum_j \tilde{\pi}_{\ell j} c_{\ell j} \right) - \sum_r \left(\sum_k \tilde{\pi}_{\ell j} c_{\ell j} \right)$$

$$= \sum_p \left(\sum_j \tilde{\pi}_{\ell j} \left(\eta_p^* + c_{\ell j} \right) \right) - \sum_r \left(\sum_k \tilde{\pi}_{\ell k} \left(\eta_r^* + c_{\ell k} \right) \right) \geq 0$$

because $\left(\eta_p^* + c_{\ell j} \right) \geq 0$ *and* $\left(\eta_r^* + c_{\ell k} \right) < 0$ *as stated at the beginning of the*
proof.

2.5 Examples of Discrete Event Optimal Supervisory Control

This section presents two examples to demonstrate different applications of
discrete-event optimal supervisors. The first example addresses health monitoring of a twin-engine unmanned aircraft that is used for surveillance and
data collection. The second example presents controlled interactions of a multiprocessor message decoding system.

2.5.1 Optimal Supervisory Control of a Twin-Engine Unmanned Aircraft

The control objective is to enhance engine safety operation. Engine health and operating conditions, which are monitored in real time based on avionic sensor information, are classified into three mutually exclusive and exhaustive categories: (i) *good*; (ii) *unhealthy (but operable)*; and (iii) *inoperable*. Upon occurrence of any observed abnormality, the supervisor decides to continue or abort the mission.

Table 2.1. Plant Automaton States

State	Description
1	Safe in base
2	Mission executing - two good engines
3	One engine unhealthy during mission execution
4	Mission executing - one good and one unhealthy engine
5	Both engines unhealthy during mission execution
6	One engine good and one engine inoperable
7	Mission execution with two unhealthy engines
8	Mission execution with only one good engine
9	One engine unhealthy and one engine inoperable
10	Mission execution with only one unhealthy engine
11	Mission aborted /not completed (Bad Marked State)
12	Mission successful (Good Marked State)
13	Aircraft destroyed (Bad Marked State)

Table 2.2. Plant Event Alphabet

Event	Event Description	Controllable Events
s	start and take-off	√
b	a good engine becoming unhealthy	
t	an unhealthy engine becoming inoperable	
v	a good engine becoming inoperable	
k	keep engine(s) running	√
a	mission abortion	√
f	mission completion	
d	destroyed aircraft	
l	landing	√

The deterministic finite state automaton model of the (unsupervised) plant (i.e., engine operation) has 13 states, of which three are marked (i.e., accepted) states, and nine events, of which four are controllable. The dump state is not included as it is not of interest in the supervisory control synthesis [8] [4]. All events are assumed to be observable. The states and events of the plant model are listed in Table 2.1 and Table 2.2, respectively. As indicated in Table 2.1, the marked states are: 11, 12 and 13, of which the states 11 and 13 are bad marked states, and the state 12 is a good marked state.

The state transition function δ (see the beginning of Section 2.2), the entries $\tilde{\pi}_{ij}$ (see Definition 2.2.4) of the event cost matrix $\tilde{\Pi}$, and the entries c_{ij} (see Definition 2.4.1) of the event disabling cost matrix C are entered simultaneously in relevant cells of Table 2.3. The dump state and any transitions to the dumped state are not shown in Table 2.3. The empty cells in Table 2.3 imply that the state transition function δ is undefined for the respective state and event. In each non-empty cell in Table 2.3, the positive integer in the first entry signifies the destination state of the transition; the non-negative fraction in the second entry is the state-based event cost $\tilde{\pi}_{ij}$; and the non-negative fraction in the third entry is the state-based event disabling cost c_{ij} of the four controllable events (i.e., events s, k, a and ℓ); event disabling cost is not applicable to the remaining five uncontrollable events (i.e., events b, t, v, f and d) and the corresponding entries are marked as "N/A". (Note that the event cost $\tilde{\pi}_{ij}$ and event disabling cost c_{ij} of a given event could be different at different states.)

The values of $\tilde{\pi}_{ij}$ were selected by extensive simulation experiments on gas turbine engine models and were also based on experience of gas turbine engine operation and maintenance. The state-based event cost $\tilde{\pi}_{ij}$ such that each row sum of the event cost matrix $\tilde{\Pi}$ is strictly less than one as given in Definition 2.2.5 and explained in detail by in a previous publication [11]. The event disabling cost c_{ij} for controllable events indicates the difficulty of disabling from the respective states and the values were chosen based on operational experience. The elements of the characteristic vector (see Definition 2.2.4) are chosen as non-negative weights based on the perception of each marked state's role on the gas turbine system performance. In this simulation example, the characteristic value of the good marked state 12 is taken to be 0.25 and those of the bad marked states 11 and 13 are taken to be –0.05 and –1.0, respectively, to quantify their respective importance; each of the remaining non-marked states is assigned zero characteristic value as seen at the bottom of Table 2.3. The information provided in Table 2.3 is sufficient to generate the state transition cost matrix Π (see Definition 2.2.7).

Based on the data given in Tables 2.1, 2.2 and 2.3, two optimal control policies - Case (a) without event disabling cost and the other Case (b) with event disabling cost have been synthesized following the respective two-step procedures in Sections 2.3 and 2.4. The results of optimal supervisor syntheses without and with event disabling cost are presented in Tables 2.4 and 2.5 supported by respective finite state machine diagrams in Figures 2.1(a) and 2.1(b). For Case(a), the event disabling cost matrix C (i.e., the relevant elements in Table 2.3) are set to zero for synthesis of the optimal control without

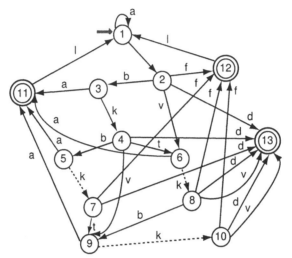

(a) Supervision without Event Disabling Cost

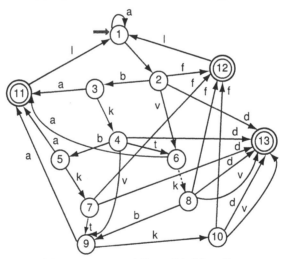

(b) Supervision with Event Disabling Cost

Fig. 2.1. Finite State Machine Diagrams of Optimally Supervised Systems

Table 2.3. State Transition δ Event Cost $\tilde{\Pi}$ and Disabling Cost C Matrices

	s	b	t	v	k	a	f	d	l
1	(2) 0.500 0.000					(1) 0.020 0.005			
2		(3) 0.050 N/A		(6) 0.010 N/A			(12) 0.800 N/A	(3) 0.100 N/A	
3					(4) 0.450 0.050	(11) 0.450 0.005			
4		(5) 0.120 N/A	(6) 0.160 N/A	(9) 0.100 N/A			(12) 0.500 N/A	(13) 0.120 N/A	
5					(7) 0.450 0.080	(11) 0.450 0.002			
6					(8) 0.450 0.010	(11) 0.450 0.004			
7			(9) 0.250 N/A				(12) 0.500 N/A	(13) 0.200 N/A	
8		(9) 0.200 N/A		(13) 0.010 N/A			(12) 0.300 N/A	(13) 0.400 N/A	
9					(10) 0.450 0.35	(11) 0.450 0.002			
10			(13) 0.350 N/A				(12) 0.200 N/A	(13) 0.400 N/A	
11									(1) 0.95 0.000
12									(1) 0.95 0.000
13									

Characteristic Vector $\bar{\chi} = [0\ 0\ 0\ 0\ 0\ 0\ 0\ 0\ 0\ -0.05\ 0.25\ -1.0]^T$
(See Definition 2.2.4)

event disabling cost. In contrast, for Case (b), all elements the event disabling cost matrix C in Table 2.3 are used for synthesis of the optimal control with event disabling cost. At successive iterations, Table 2.4 lists the performance

Table 2.4. Synthesis without Event Disabling Cost

Iteration 0	Iteration 1	Iteration 2
0.0823	0.0840	0.0850
0.1613	0.1646	0.1665
0.0062	0.0134	0.0366
-0.0145	0.0500	0.0506
-0.0367	0.0134	0.0138
-0.1541	0.0134	0.0138
-0.1097	-0.0317	-0.0312
-0.3706	-0.3084	-0.3080
-0.2953	0.0134	0.0138
-0.6844	-0.6840	-0.6839
0.0282	0.0298	0.0307
0.3282	0.3298	0.3307
-1.0000	-1.0000	-1.0000

Table 2.5. Synthesis with Event Disabling Cost

Iteration 0	Iteration 1	Iteration 2
0.0823	0.0839	0.0841
0.1613	0.1645	0.1649
0.0062	0.0134	0.0188
-0.0145	0.0117	0.0118
-0.0367	-0.0356	-0.0354
-0.1541	0.0034	0.0035
-0.1097	-0.1088	-0.1086
-0.3706	-0.3700	-0.3699
-0.2953	-0.2944	-0.2943
-0.6844	-0.6841	-0.6840
0.0282	0.0297	0.0299
0.3282	0.3297	0.3299
-1.0000	-1.0000	-1.0000

vectors in Case (a): $\bar{\mu}^0$ for the unsupervised (i.e., open loop) plant, $\bar{\mu}^1$ in iteration 1, and $\bar{\mu}^2$ in iteration 2 when the synthesis is completed because of no sign change between elements of $\bar{\mu}^1$ and $\bar{\mu}^2$. Table 2.4 shows that $\bar{\mu}^2 \geq \bar{\mu}^1 \geq \bar{\mu}^0$ elementwise. This is due to disabling the controllable event k leading to states 7, 8 and 10 as indicated by the dashed arcs in the state transition diagram of Figure 2.1(a). Consequently, the states 7, 8, and 10 become isolated as there are no other events leading to these states. Starting with the initial state 1, indicated by an external arrow in Figure 2.1(a), the optimal performance is 0.0850 that is the first element μ_1^2 of the performance vector $\bar{\mu}^2$ as seen in the top right hand corner in Table 2.4.

The results are different for Case (b) because the event disabling cost is taken into account in optimal supervisor synthesis as seen in Table 2.5 and Figure 2.1(b); in this case, only the state 8 is isolated due to disabling of the controllable event k at the state 6. At successive iterations, Table 2.5 lists the performance vectors for this Case (b) where $\bar{\eta}^0 = \bar{\mu}^0$ for the unsupervised (i.e., open loop) plant; $\bar{\eta}^1$ in iteration 1, and $\bar{\eta}^2$ in iteration 2 when the synthesis is completed because of no sign change between elements of $\bar{\eta}^1$ and $\bar{\eta}^2$. (Note that, in general, the number of iterations needed for supervisor synthesis without and with event disabling cost may not be the same.) Table 2.5 shows that $\bar{\eta}^2 \geq \bar{\eta}^1 \geq \bar{\eta}^0$ elementwise. This is due to disabling of the controllable event k leading to the state 8 as indicated by the dashed arcs in the state transition diagram of Figure 2.1(b). Consequently, the state 8 (shown in a dotted circle in Figure 2.1 (b)) becomes isolated as there are no other events leading to this state. Starting with the initial state 1, indicated by an arrow in Figure 2.1(b), the optimal performance is 0.0841 that is the first element $\bar{\eta}_1^2$ of the performance vector $\bar{\eta}^2$ as seen in the top right hand corner in Table 2.5. Clearly, the performance of the supervisor in Case (b) is suboptimal with respect to that in Case (a). That is, the performance in Case (b) cannot excel the performance in Case (a)) where the event disabling cost is not taken into account.

2.5.2 Optimal Supervisory Control of a Three-Processor Message Decoding System

This section presents the design of a discrete-event (controllable) supervisor for a multiprocessor message decoding system as described below.

Figure 2.2(a) depicts the arrangement of the message decoding system, where each of the three processors, p_1, p_2 and p_3, receives encoded message to be decoded. The processor p_3 normally receives the most important messages, and p_1 receives the least important messages. There is a server between each pair of processors - s_1 between p_1 and p_2; s_2 between p_2 and p_3; and s_3 between p_3 and p_1. Each server is connected to each of its two adjacent processors by a link - the server s_j is connected to the adjacent processors p_i and p_k through the links L_{ij} and L_{kj}, respectively. Out of these six links, each of the three links, L_{11}, L_{12}, and L_{21}, is equipped with a switch to disable the respective connection whenever it is necessary; each of the remaining three links, L_{22}, L_{32}, and L_{33}, always remain connected. Each server s_i is equipped with a decoding key k_i that, at any given time, can only be accessed by only one of the two processors, adjacent to the server, through the link connecting the processor and the server. In order to decode the message, the processor holds the information on both keys of the servers next to it, one at a time. After decoding, the processor simultaneously releases both keys so that other processors may get hold of them.

The unsupervised plant model of the decoding system is depicted as a finite state machine in Figure 2.2(b), where state 1 is the initial state; the states

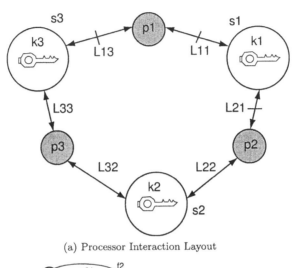

(a) Processor Interaction Layout

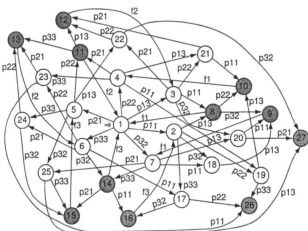

(b) DFSA Model of Unsupervised Processor Interactions

Fig. 2.2. Three-processor Decoding System

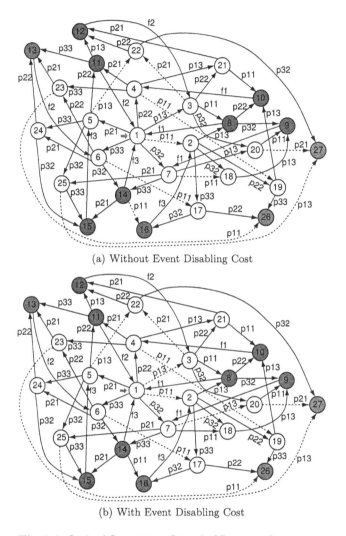

(a) Without Event Disabling Cost

(b) With Event Disabling Cost

Fig. 2.3. Optimal Supervisory Control of Processor Interactions

Table 2.6. Event Cost Matrix $\tilde{\Pi}$ and Characteristic Vector $\tilde{\chi}$

	p11 (1)	p13 (2)	p21 (3)	p22 (4)	p32 (5)	p33 (6)	f1 (7)	f2 (8)	f3 (9)	X
1	0.15	0.11	0.15	0.15	0.15	0.3				0
2		0.1		0.25	0.25	0.3				0
3	0.2		0.2	0.2	0.2					0
4	0.25	0.1	0.25			0.3				0
5		0.1		0.25	0.25	0.3				0
6	0.2		0.2	0.2	0.2					0
7	0.25	0.1	0.25			0.3				0
8				0.35	0.35		0.2			0.01
9							0.5			0.01
10							0.5			0.01
11		0.1				0.3		0.3		0.02
12								0.5		0.02
13								0.5		0.02
14	0.3		0.3						0.3	0.04
15									0.4	0.04
16									0.4	0.04
17				0.4	0.4					0
18		0.1				0.3				0
19		0.1				0.3				0
20	0.4		0.4							0
21	0.4		0.4							0
22				0.4	0.4					0
23	0.4		0.4							0
24				0.4	0.4					0
25		0.1				0.3				0
26										-1
27										-1

26 and 27, shaded in red, are bad marked states representing deadlocks; the states, 11, 12, 13, 14, 15 and 16, shaded in gray, are good marked states; and the remaining unshaded states are unmarked. The event p_{ij} indicates that processor p_i has accessed the key k_j; and the event f_i indicates that the processor p_i has finished decoding and (simultaneously) released both keys in its possession upon completion of decoding. The events f_i are uncontrollable because, after the decoding is initiated, there is no control on when a processor finishes decoding. Table 2.6 lists the event cost matrix $\tilde{\Pi}$ and the characteristic vector $\bar{\chi}$. In the right hand column of Table 2.6, positive values are assigned to the χ values of the states 8 to 16 that represent successful decoding of each processor. The χ values of the deadlock states 26 and 27 (where each processor holds exactly one key and hence no processor releases its key) are assigned the negative value of -1. The remaining states are non-marked and are assigned zero χ values. Table 2.7 lists the event disabling cost matrix C where the

Table 2.7. Event Disabling Cost Matrix C

	$p11$ (1)	$p13$ (2)	$p21$ (3)	$p22$ (4)	$p32$ (5)	$p33$ (6)	$f1$ (7)	$f2$ (8)	$f3$ (9)
1	0.1	0.06	0.0333	N/A	N/A	N/A	N/A	N/A	N/A
2	N/A	0.06	N/A	N/A	N/A	N/A	N/A	N/A	N/A
3	0.075	N/A	0.025	N/A	N/A	N/A	N/A	N/A	N/A
4	0.06	0.06	0.02	N/A	N/A	N/A	N/A	N/A	N/A
5	N/A	0.06	N/A	N/A	N/A	N/A	N/A	N/A	N/A
6	0.075	N/A	0.02	N/A	N/A	N/A	N/A	N/A	N/A
7	N/A	N/A	N/A	N/A	N/A	N/A	N/A	N/A	N/A
8	N/A	N/A	N/A	N/A	N/A	N/A	N/A	N/A	N/A
9	N/A	N/A	N/A	N/A	N/A	N/A	N/A	N/A	N/A
10	N/A	N/A	N/A	N/A	N/A	N/A	N/A	N/A	N/A
11	N/A	0.06	N/A	N/A	N/A	N/A	N/A	N/A	N/A
12	N/A	N/A	N/A	N/A	N/A	N/A	N/A	N/A	N/A
13	N/A	N/A	N/A	N/A	N/A	N/A	N/A	N/A	N/A
14	0.05	N/A	0.01667	N/A	N/A	N/A	N/A	N/A	N/A
15	N/A	N/A	N/A	N/A	N/A	N/A	N/A	N/A	N/A
16	N/A	N/A	N/A	N/A	N/A	N/A	N/A	N/A	N/A
17	N/A	N/A	N/A	N/A	N/A	N/A	N/A	N/A	N/A
18	N/A	0.06	N/A	N/A	N/A	N/A	N/A	N/A	N/A
19	N/A	0.06	N/A	N/A	N/A	N/A	N/A	N/A	N/A
20	0.0375	N/A	0.0125	N/A	N/A	N/A	N/A	N/A	N/A
21	0.0375	N/A	0.0125	N/A	N/A	N/A	N/A	N/A	N/A
22	N/A	N/A	N/A	N/A	N/A	N/A	N/A	N/A	N/A
23	0.0375	N/A	0.0125	N/A	N/A	N/A	N/A	N/A	N/A
24	N/A	N/A	N/A	N/A	N/A	N/A	N/A	N/A	N/A
25	N/A	0.06	N/A	N/A	N/A	N/A	N/A	N/A	N/A
26	N/A	N/A	N/A	N/A	N/A	N/A	N/A	N/A	N/A
27	N/A	N/A	N/A	N/A	N/A	N/A	N/A	N/A	N/A

entries c_{ij} are non-negative fractions in the first three columns corresponding to the controllable events, p_{11}, p_{13} and p_{21}. These disabling costs are assigned in accordance with the characteristic values of the respective marked states. The entries, marked as "N/A" correspond to either uncontrollable events or the states at which the plant model does not generate a controllable event.

Based on the optimal control policies described in Sections 2.3 and 2.4, two supervisors have been synthesized without and with event disabling cost, as shown in Figures 2.3(a) and 2.3(b), respectively. A comparison of these two diagrams reveals that different controllable events are disabled at different states, as indicated by arcs with dashed lines. Specifically, controllable events, causing state transitions $1 \rightarrow 2$, $3 \rightarrow 22$, $4 \rightarrow 19$, $5 \rightarrow 22$, $6 \rightarrow 17$, $7 \rightarrow 18$, $14 \rightarrow 16$, $20 \rightarrow 27$, $23 \rightarrow 26$, and $25 \rightarrow 27$, are disabled under the control policy without event disabling cost, as seen in Figure 2.3(a). In contrast,

Table 2.8. Performance without Event Disabling Cost

States	Iteration 0	Iteration 1	Iteration 2
1	-0.1646	0.059	0.0161
2	-0.2141	-0.2030	-0.1991
3	-0.1970	0	0.0104
4	-0.2277	0	0.0145
5	-0.0902	0.0198	0.0223
6	-0.1788	0.0112	0.0207
7	-0.0765	0.0168	0.0224
8	-0.0692	0.0211	0.0267
9	-0.0283	0.0184	0.0212
10	-0.1039	0.0100	0.0172
11	-0.0581	0.0295	0.0365
12	-0.0785	0.0200	0.0252
13	-0.0694	0.0256	0.0303
14	-0.0219	0.0561	0.0595
15	0.0039	0.0479	0.0489
16	-0.0456	-0.0412	-0.0396
17	-0.4183	-0.4165	-0.4159
18	-0.0165	-0.0124	-0.0098
19	-0.3104	-0.3000	-0.2983
20	-0.4113	0	0.0085
21	-0.0729	0	0.0170
22	-0.4134	-0.3920	-0.3899
23	-0.4278	0	0.0121
24	-0.0262	0.0294	0.0317
25	-0.0988	0.0144	0.0147
26	-1.0000	-1.0000	-1.0000
27	-1.0000	-1.0000	-1.0000

controllable events, causing state transitions $1 \to 2$, $1 \to 3$, $3 \to 22$, $4 \to 19$, $5 \to 22$, $6 \to 17$, $7 \to 20$, $20 \to 27$, $23 \to 26$, and $25 \to 27$, are disabled under the the control policy with event disabling cost, as seen in Figure 2.3(b). The rationale for the difference in event disabling in these two cases is that extremal point(s) on the optimization surface shift due to introduction of non-zero disabling costs in the supervisor synthesis algorithms as described in Sections 2.3.1 and 2.4.1.

Tables 2.8 and 2.9 show the corresponding performance vectors for optimal control without and with event disabling at each iteration; the optimization is terminated after iterations 2 and 4, respectively, when optimality is reached and the performance cannot be improved any further by event disabling or re-enabling. The performance is non-decreasing at every state from one iteration to the next as seen in Tables 2.8 and 2.9. Since signs of the performance vector elements have changed for several states at iteration 1 in Table 2.8, the procedure is continued to iteration 2 where it is terminated because of no

Table 2.9. Performance with Event Disabling Cost

States	Iteration 0	Iteration 1	Iteration 2	Iteration 3	Iteration 4
1	-0.1646	-0.0364	-0.0228	-0.0228	-0.0216
2	-0.2141	-0.2110	-0.2032	-0.2015	-0.2012
3	-0.1970	-0.0194	-0.0045	-0.0011	-0.0005
4	-0.2277	-0.0270	-0.0134	-0.0133	-0.0061
5	-0.0902	0.0034	0.0081	0.0081	0.0082
6	-0.1788	-0.0184	-0.0050	-0.0049	-0.0049
7	-0.0765	-0.0057	0.0031	0.0032	0.0033
8	-0.0692	0.0040	0.0106	0.0107	0.0122
9	-0.0283	0.0072	0.0115	0.0116	0.0117
10	-0.1039	-0.0035	0.0033	0.0033	0.0069
11	-0.0581	0.0063	0.0202	0.0204	0.0208
12	-0.0785	0.0103	0.0177	0.0194	0.0198
13	-0.0694	0.0108	0.0175	0.0175	0.0176
14	-0.0219	0.0282	0.0337	0.0340	0.0343
15	0.0039	0.0414	0.0432	0.0433	0.0433
16	-0.0456	-0.0444	-0.0413	-0.0406	-0.0405
17	-0.4183	-0.4178	-0.4165	-0.4162	-0.4162
18	-0.0165	-0.0126	-0.0112	-0.0110	-0.0110
19	-0.3104	-0.3060	-0.3060	-0.2997	-0.2993
20	-0.4113	-0.0021	-0.0004	-0.0004	-0.0003
21	-0.0729	-0.0200	-0.0079	0.0091	0.0107
22	-0.4134	-0.3959	-0.3929	-0.3922	-0.3921
23	-0.4278	-0.0200	-0.0080	-0.0080	-0.0080
24	-0.0262	0.0209	0.0243	0.0243	0.0243
25	-0.0988	0.0064	0.0070	0.0070	0.0070
26	-1.0000	-1.0000	-1.0000	-1.0000	-1.0000
27	-1.0000	-1.0000	-1.0000	-1.0000	-1.0000

further sign change. Similarly, in Table 2.9, sign change in the performance vector element(s) continues until iteration 3 and the procedure is terminated at iteration 4 because of no further sign change. It is also seen that, starting from the same condition at iteration 0 for the unsupervised plant, the performance of the supervisor with event disabling cost is always inferior to that without event disabling cost because of the additional penalty. For example, with starting state 1, the performance of the supervised plant without event disabling cost is 0.0161 and that with event disabling cost is -0.0216 while the performance of the unsupervised plant is -0.1646.

2.6 Summary and Conclusions

This chapter presents the theory, formulation, and validation of optimal supervisory control policies for dynamical systems, modeled as deterministic finite state automata (DFSA), which may have already been subjected to constraints such as control specifications. The synthesis procedure for optimal control without and with event disabling cost is quantitative and relies on a signed real measure of regular languages, which is based on a specified state transition cost matrix and a characteristic vector [12] [9] [11].

The state-based optimal control policy without event disabling cost maximizes the language measure vector $\bar{\mu}$ by attempting to selectively disable controllable events that may lead to bad marked states and simultaneously ensuring that the remaining controllable events are kept enabled. The goal is to maximize the measure of the controlled plant language without any further constraints. The control policy induced by the updated state transition cost matrix yields maximal performance and is unique in the sense that the controlled language is most permissive (i.e., least restrictive) among all controller(s) having the optimal performance.

The performance measure vector $\bar{\eta}$, for optimal control with disabling cost, is obtained as the language measure vector of the supervised plant minus the disabling cost characteristic vector. The optimal control policy maximizes the performance vector elementwise by attempting to avoid termination on bad marked states by selectively disabling controllable events with reasonable disabling costs, and simultaneously ensuring that the remaining controllable events are kept enabled. As the cost of event disabling approaches zero, the optimal control policy with event disabling cost converges to that without event disabling cost.

Derivation of the optimal supervisory control policies requires at most n iterations, where n is the number of states of the DFSA model and each iteration is required to solve a set of n simultaneous linear algebraic equations having complexity of $O(n^3)$ [9] [11]. As such computational complexity of the control synthesis procedure is polynomial in the number of DFSA model states. The procedure for synthesis of the optimal control algorithm has been validated on the DFSA model of a twin-engine surveillance aircraft and the DFSA model of a three-processor decoding operation.

Future areas of research in optimal control include robustness of the control policy relative to unstructured and structured uncertainties in the plant model including variations in the language measure parameters [2].

References

1. Y. Brave and M. Heyman, *On optimal attraction of discrete-event processes*, Information Science **67** (1993), no. 3, 245–276.
2. J. Fu, C.M. Lagoa, and A. Ray, *Robust optimal control of regular languages with event cost uncertainties*, Proceedings of IEEE Conference on Decision and Control, December 2003, pp. 3209–3214.

3. J. Fu, A. Ray, and C.M. Lagoa, *Optimal control of regular languages with event disabling cost*, Proceedings of American Control Conference, Denver, Colorado, June 2003, pp. 1691–1695.
4. J. Fu, A. Ray, and C.M. Lagoa, *Unconstrained optimal control of regular languages*, Automatica **40** (2004), no. 4, 639–648.
5. R. Kumar and V. Garg, *Optimal supervisory control of discrete event dynamical systems*, SIAM J. Control and Optimization **33** (1995), no. 2, 419–439.
6. A.W. Naylor and G.R. Sell, *Linear operator theory in engineering and science*, Springer-Verlag, 1982.
7. K. Passino and P. Antsaklis, *On the optimal control of discrete event systems*, Proceedings of 28th IEEE Decision and Control Conference, Tampa, Florida (1989), 2713–2718.
8. P.J. Ramadge and W.M. Wonham, *Supervisory control of a class of discrete event processes*, SIAM J. Control and Optimization **25** (1987), no. 1, 206–230.
9. A. Ray and S. Phoha, *Signed real measure of regular languages for discrete-event automata*, Int. J. Control **76** (2003), no. 18, 1800–1808.
10. R. Sengupta and S. Lafortune, *An optimal control theory for discrete event systems*, SIAM J. Control and Optimization **36** (1998), no. 2, 488–541.
11. A. Surana and A. Ray, *Signed real measure of regular languages*, Demonstratio Mathematica **37** (2004), no. 2, 485–503.
12. X. Wang and A. Ray, *A language measure for performance evaluation of discrete-event supervisory control systems*, Applied Mathematical Modelling **28** (2004), no. 9, 817–833.

3

Robust Optimal Control of Regular Languages

Constantino Lagoa[1], Jinbo Fu[2], and Asok Ray[3]

[1] The Pennsylvania State University, University Park, PA 16802
   lagoa@engr.psu.edu
[2] The Pennsylvania State University, University Park, PA 16802 jxf293@psu.edu
[3] The Pennsylvania State University, University Park, PA 16802 axr2@psu.edu

Summary. This chapter presents an algorithm for robust optimal control of regular languages under specified uncertainty bounds for the event costs of a language measure that has been recently reported in literature and is presented in Chapter 1. The performance index for the proposed robust optimal policy is obtained by combining the measure of the supervised plant language with uncertainty. The performance of a controller is represented by the language measure of the supervised plant and is minimized over the given range of event cost uncertainties. Synthesis of the robust optimal control policy requires at most n iterations, where n is the number of states of the deterministic finite state automaton (DFSA) model generated from the regular language of the unsupervised plant behavior. The computational complexity of control synthesis is polynomial in n.

Key words: Discrete Event Supervisory Control, Optimal Control, Robust Control, Language Measure

3.1 Introduction

This chapter addresses the problem of robust supervisory control of regular languages, representing deterministic finite-state automata (DFSA) models of the physical plants under decision and control. Specifically, algorithms are formulated for both robust analysis and optimal robust supervisor synthesis for regular languages, or equivalently, for their DFSA models. Also, mathematical foundations of these algorithms are rigorously established.

Recent results on the analysis of DFSA and on the synthesis of optimal supervisory control policies motivate the work presented in this chapter. More precisely, a novel way of accessing the performance of DFSA has been proposed in [16] [14] [12], where a signed real measure of regular languages has been developed. This measure is computed using an event cost matrix and a characteristic vector and it provides a quantitative mean for evaluating the performance of the regular language that represents the discrete-event dynamic behavior of the DFSA plant model. This work was followed by [5] where, based on this performance measure, an algorithm is developed for the design of optimal supervisors. The optimal supervisory controller is obtained by selectively disabling controllable events in order to maximize the overall performance. However, uncertainty was not addressed. This chapter considers structural uncertainty in the DFSA model: (i) uncertainty in the presence of some of the state transitions; and (ii) uncertainty in the entries of the event cost matrix (and hence the state transition matrix) used to compute the performance of the system. The first source of uncertainty results from inaccuracies in modelling of the discrete-event dynamic behavior of the plant under supervision. The second source of uncertainty is inherent to the process of parameter identification of the event cost matrix. Since the entries of the event cost matrix are often determined by Monte Carlo simulations and, hence, their values are only known as bounded intervals with a specified level of statistical confidence. This issue of uncertainty has been addressed in Chapter 1 and is further discussed in this chapter from the perspectives of robust optimal supervisor synthesis. The robust supervisory control algorithm, presented in this chapter, has a complexity that is polynomial in the number of the states of the DFSA model.

3.1.1 Previous Work

The problem of robust control of discrete-event dynamical systems (DEDS) has been addressed by several researchers. Park and Lim [9] have studied the problem of robust control of nondeterministic DEDS. The performance measure used was nonblocking property of the supervised plant. Necessary and sufficient conditions for the existence of a robust nonblocking controller for a given finite set of plants are provided. However, no algorithm for controller design is provided. The problem of designing nonblocking robust controllers was also addressed by Cury and Krogh [2] with the additional constraint that the infinite behavior belongs to a given set of allowable behaviors. In this work, the authors concentrated on the problem of designing a controller that maximizes the set of plants for which their supervised behavior belongs to the admissible set of behaviors. Takai [15] addresses a similar problem. However, it considers the whole behavior (not just the infinite behavior) and it does not consider nonblockingness. Lin [7] adopted a different approach, where both the set of admissible plants and the performance are defined in terms of the marked language. Taking the set of admissible plants as the plants whose marked language is in between two given behaviors, the authors provided conditions for solving the problem of controller design such that the supervised behavior of an admissible plant model contains a desired behavior K.

To address a subject related to that of this chapter and the previous one, several researchers have proposed optimal control algorithms for deterministic finite state automata (DFSA) based on different assumptions. Some of these researchers have attempted to quantify the controller performance using different types of cost assigned to the individual events. Passino and Antsaklis [10] proposed path costs associated with state transitions and hence optimal control of a discrete event system is equivalent to following the shortest path on the graph representing the uncontrolled system. Kumar and Garg [6] introduced the concept of payoff and control costs that are incurred only once regardless of the number of times the system visits the state associated with the cost. Consequently, the resulting cost is not a function of the dynamic behavior of the plant. Brave and Heymann [1] introduced the concept of optimal attractors in discrete-event control. Sengupta and Lafortune [13] used control cost in addition to the path cost in optimization of the performance index for trade-off between finding the shortest path and reducing the control cost.

A limitation of the work mentioned above is that the controllers are designed so that the closed loop system has some characteristics. No performance measure is given that can compare the performance of different controllers. To address this issue Wang and Ray [16], followed by Ray and Phoha [12], have proposed a signed measure for regular languages. This novel way of addressing the performance of DFSAs enable the development of a novel approach to supervisor design. The design of optimal supervisor has been reported for without and with event disabling cost by Fu, Ray and Lagoa in [5] and [4], respectively, and is also described in Chapter 2. Apparently, the impact of uncertainty on the synthesis of discrete-event supervisory control systems is first reported by Fu, Lagoa and Ray [3]. This chapter extends the work on optimal control, described in Chapter 2, by including robustness in the problem of designing supervisors in the presence of uncertainty.

3.2 An Alternative Approach to Performance Computation

This section provides an alternative way of optimizing the performance measure of regular languages to the one presented in Chapter 2. Before this is done, a brief review of language measure is presented for the sake of completeness.

3.2.1 Brief Review of the Language Measure

This section briefly reviews the concept of signed real measure of regular languages introduced in Chapter 1. As before, let the plant behavior be modelled as a DFSA $G_i \equiv (Q, \Sigma, \delta, q_i, Q_m)$ where Q is the finite set of states with $|Q| = n$ excluding the dump state [11] if any, and $q_i \in Q$ is the initial state; Σ is the (finite) alphabet of events; Σ^* is the set of all finite-length strings of events including the empty string ϵ; the (total) function $\delta : Q \times \Sigma \to Q$ represents state transitions and $\hat{\delta} : Q \times \Sigma^* \to Q$ is an extension of δ; and $Q_m \subseteq Q$ is the set of marked states.

The set Q_m is partitioned into Q_m^+ and Q_m^-, where Q_m^+ contains all *good* marked states that are desired to terminate on and Q_m^- contains all *bad* marked states that are not desired to terminate on, although it may not always be possible to avoid the bad states while attempting to reach the good states. The marked language $L_m(G_i)$, associated with DFSA G_i, is partitioned into $L_m^+(G_i)$ and $L_m^-(G_i)$ consisting of good and bad strings that, starting from q_i, terminate on Q_m^+ and Q_m^-, respectively.

The language of all strings that, starting at a state $q_i \in Q$, terminates on a state $q_j \in Q$, is denoted as $L_{i,j}$. That is, $L_{i,j} \equiv \{s \in L(G_i) : \hat{\delta}(q_i, s) = q_j\}$. Furthermore, a characteristic function is assumed to exist which assigns a signed real weight to state partitioned sublanguages $L_{i,j}$. This function is denoted by $\chi : Q \to [-1, 1]$ and is such that

$$\chi(q_j) \in \begin{cases} (0, 1] & \text{if } q_j \in Q_m^+ \\ \{0\} & \text{if } q_j \notin Q_m \\ [-1, 0) & \text{if } q_j \in Q_m^- \end{cases}$$

Now, the event cost is defined as $\tilde{\pi} : \Sigma^* \times Q \to [0,1)$ such that $\forall q_j \in Q$, $\forall \sigma_k \in \Sigma$, $\forall s \in \Sigma^*$,

- $\tilde{\pi}[\sigma_k|q_j] = 0$ if $\delta(q_j, \sigma_k)$ is undefined; $\tilde{\pi}[\epsilon|q_j] = 1$;
- $\tilde{\pi}[\sigma_k|q_j] \equiv \tilde{\pi}_{jk} \in [0,1)$; $\sum_k \tilde{\pi}_{jk} < 1$;
- $\tilde{\pi}[\sigma_k \ s|q_j] = \tilde{\pi}[\sigma_k|q_j]\tilde{\pi}[s|\delta(q_j, \sigma_k)]$.

Now, as before, the language measure is defined incrementally. The signed real measure μ of a singleton string set $\{s\} \subset L_{i,j} \subseteq L(G_i) \in 2^{\Sigma^*}$ is defined as:

$$\mu(\{s\}) \equiv \chi(q_j)\,\tilde{\pi}(s|q_i) \ \forall s \in L_{i,j}.$$

The signed real measure of $L_{i,j}$ is defined as

$$\mu(L_{i,j}) \equiv \sum_{s \in L_{i,j}} \mu(\{s\})$$

and the signed real measure of a DFSA G_i, initialized at the state $q_i \in Q$, is denoted as:

$$\mu_i \equiv \mu(L(G_i)) = \sum_j \mu(L_{i,j})$$

To provide an efficient way of computing the language measure defined above, one needs the concept of *state transition cost matrix*. To this end, first define the state transition cost of the DFSA. This cost is defined as a function $\pi : Q \times Q \to [0,1)$ such that

$$\forall q_j, q_k \in Q, \quad \pi(q_k|q_j) = \sum_{\sigma \in \Sigma : \delta(q_j, \sigma) = q_k} \tilde{\pi}(\sigma|q_j) \equiv \pi_{jk}$$

and

$$\pi_{jk} = 0 \text{ if } \{\sigma \in \Sigma : \delta(q_j, \sigma) = q_k\} = \emptyset.$$

The $n \times n$ state transition cost matrix, denoted as Π-matrix, is defined as:

$$\Pi = \begin{bmatrix} \pi_{11} & \pi_{12} & \cdots & \pi_{1n} \\ \pi_{21} & \pi_{22} & \cdots & \pi_{2n} \\ \vdots & & \ddots & \vdots \\ \pi_{n1} & \pi_{n2} & \cdots & \pi_{nn} \end{bmatrix}.$$

where each row sum in the Π-matrix is strictly less than 1, i.e., $\exists\, \theta \in (0,1]$ such that

$$\sum_j \pi_{ij} \leq (1 - \theta) \ \forall i$$

An efficient way of computing the language measure above is now provided. Wang and Ray [16] have shown that the measure $\mu_i \equiv \mu(L(G_i))$ of the language $L(G_i)$ can be expressed as:

$$\mu_i = \sum_j \pi_{ij}\mu_j + \chi_i$$

where $\chi_i \equiv \chi(q_i)$. Equivalently, in vector notation:

$$\mu = \Pi\mu + X$$

where the measure vector $\mu \equiv [\mu_1 \ \mu_2 \ \cdots \ \mu_n]^T$ and the characteristic vector $X \equiv [\chi_1 \ \chi_2 \ \cdots \ \chi_n]^T$. Hence, since by construction the matrix $I - \Pi$ is invertible, the measure vector μ is uniquely determined as

$$\mu = [I - \Pi]^{-1}X.$$

3.2.2 A New Look on Computing Performance

As seen above, for a given automaton and its associated event cost matrix Π, the problem of computing performance has been formulated as the solution of a set of linear equations, i.e., computing performance is equivalent to finding a performance vector μ satisfying the following linear equalities

$$\mu = \Pi\mu + X$$

where X is the characteristic vector and Π is the event cost matrix. Although very useful for designing optimal controllers, this kind of approach has limited applicability to uncertain systems.

To address the problem of robust controller design, this chapter takes a different look at the problem of computing performance. For a given automata with n states and an associated event cost matrix Π, define the function (or operator) $T : \mathbf{R}^n \to \mathbf{R}^n$

$$T(x) \equiv \Pi x + X.$$

Now, the problem of finding the performance of the automaton is equivalent of finding the fixed point of T, i.e., μ is the performance vector if and only if it satisfies the equality

$$\mu = T(\mu).$$

Although it seems that one is replacing a problem by another of equivalent complexity, it turns out that the operator T is a contraction mapping and many well known results on contraction mapping theory can be made use of. This will become apparent in the later part of the chapter where algorithms for robustness analysis and robust controller design are provided. Another advantage of this approach, although not explored in this chapter, is that it provides a way of addressing the problem of analysis and controller design for non regular languages. That is, using an operator viewpoint, one can address the problem of controlling a discrete event system with an infinite number of states.

3.3 Modelling Uncertainty

This section introduces the model of uncertainty of a DFSA along with the concept of robust performance. Uncertainty in the event cost matrix and in the existence of state transitions are the types of uncertainty studied in this chapter. Other types of uncertainties (e.g., those due to the number of states and controllability of the events) that are not addressed here are topics of future work.

3.3.1 Uncertainty

As mentioned above, two types of uncertainties are considered in the model used in this chapter: Uncertainty in the presence of state transitions and uncertainty in the event cost matrix. These two sources of uncertainty can be modelled by modifying the definition of the measure used to compute the performance. More precisely, define the *uncertain event cost* as

$$\tilde{\pi}^{\Delta}[\sigma_j|q_i] = \left(\tilde{\pi}^0[\sigma_j|q_i] + \Delta_{cost}[\sigma_j|q_i]\right)\Delta_{model}[\sigma_j|q_i]$$

where $\tilde{\pi}^0[\sigma_j|q_i]$ is the event cost of the nominal model, $\Delta_{cost}[\sigma_j|q_i]$ represents the uncertainty associated with the determination of the event cost and belongs to an interval that is a proper subset of $[0,1]$. Finally, $\Delta_{model}[\sigma_j|q_i]$ represents uncertainty in the existence of this specific transition of the automaton; i.e., if this transition is always present then

$$\Delta_{model}[\sigma_j|q_i] = 1.$$

If there is uncertainty in the presence of this transition, then

$$\Delta_{model}[\sigma_j|q_i] \in \{0,1\}.$$

Furthermore, it is assumed that the uncertain event costs vary independently; i.e., each uncertain parameter only enters in one event cost. Finally, it is also assumed that there exists constant $\theta > 0$ such that for all admissible uncertainty values and all i

$$\sum_j \sum_{\sigma \in \Sigma: \delta(q_i,\sigma)=q_j} \tilde{\pi}^{\Delta}[\sigma|q_i] \leq 1 - \theta.$$

The set of admissible values for the uncertainty Δ is denoted by $\boldsymbol{\Delta}$ and is a compact subset of $[0,1]^{|\Sigma| \times n} \times \{0,1\}^{|\Sigma| \times n}$. Now, given any $\Delta \in \boldsymbol{\Delta}$, the uncertain state transition matrix is given by

$$\Pi(\Delta) = \begin{bmatrix} \pi_{11}(\Delta) & \pi_{12}(\Delta) & \dots & \pi_{1n}(\Delta) \\ \pi_{21}(\Delta) & \pi_{22}(\Delta) & \dots & \pi_{2n}(\Delta) \\ \vdots & \vdots & \ddots & \vdots \\ \pi_{n1}(\Delta) & \pi_{n2}(\Delta) & \dots & \pi_{nn}(\Delta) \end{bmatrix}$$

where

$$\pi_{ij}(\Delta) = \sum_{\sigma \in \Sigma : \delta(q_i, \sigma) = q_j} \tilde{\pi}^{\Delta}[\sigma | q_i].$$

3.3.2 Additional Notation

Given a supervisor S, let $\Pi(S, \Delta)$ be the uncertain state transition matrix under supervisor S; i.e., $\Pi(S, \Delta)$ has entries

$$\pi_{ij}(S, \Delta) = \sum_{\sigma \in \Sigma : \delta(q_i, \sigma) = q_j} \tilde{\pi}^{S, \Delta}[\sigma | q_i].$$

where

$$\tilde{\pi}^{S, \Delta}[\sigma | q_i] = \begin{cases} 0 & \text{if event } \sigma \text{ is controllable and is disabled by } S; \\ \tilde{\pi}^{\Delta}[\sigma | q_i] & \text{otherwise.} \end{cases}$$

For a given admissible value for the uncertainty Δ, the performance of the plant under the supervisor S, denoted by $\mu(S, \Delta)$, is the solution of

$$\mu(S, \Delta) = \Pi(S, \Delta)\mu(S, \Delta) + X.$$

3.4 Robust Performance

In this section, a precise definition of robust performance is provided and an algorithm for computing is presented.

3.4.1 Definition of Robust Performance

Consider an uncertain automaton controlled by a supervisor S. The *robust performance of supervisor* S, denoted by $\underline{\mu}(S)$ is defined as the worst-case performance; i.e.,

$$\underline{\mu}(S) = \min_{\Delta \in \mathbf{\Delta}} \mu(S, \Delta)$$

where the minimum above is taken elementwise. Although the minimization is done element by element, this performance is achieved for some $\Delta^* \in \mathbf{\Delta}$. The precise statement of this result is given below.

Lemma 3.4.1 *Let S be a supervisor. Then, there exists a $\Delta^* \in \mathbf{\Delta}$ such that, for all admissible $\Delta \in \mathbf{\Delta}$,*

$$\underline{\mu}(S) = \mu(S, \Delta^*) \leq \mu(S, \Delta)$$

where the inequality above is an elementwise one.

Proof. See Section 3.5.2.

An algorithm for computing $\underline{\mu}(S)$ is now provided. This algorithm is a generalization of the original work, reported in [5], where a similar algorithm was provided for the case of optimal controller design.

Algorithm 3.4.1 *Computation of worst–case performance of supervisor S.*

Step 0. Let $k = 0$ and pick $\Delta^0 \in \Delta$.
Step 1. Let Δ^{k+1} be such that[4]

$$\pi_{ij}(S, \Delta^{k+1}) = \begin{cases} \max_{\Delta \in \Delta} \pi_{ij}(S, \Delta) & \text{if } \mu_j(S, \Delta^k) < 0; \\ \min_{\Delta \in \Delta} \pi_{ij}(S, \Delta) & \text{if } \mu_j(S, \Delta^k) \geq 0. \end{cases}$$

Step 2. If $\underline{\mu}(S, \Delta^{k+1}) = \underline{\mu}(S, \Delta^k)$ then $\underline{\mu}(S) = \underline{\mu}(S, \Delta^{k+1})$ and stop. Else let $k \leftarrow k + 1$ and go to Step 1

The theorem below presents the formal result that indicates that the above algorithm converges in a finite number of steps.

Theorem 3.4.1 *Given a supervisor S, the above algorithm converges to its robust performance; i.e.,*

$$\mu(S, \Delta^k) \rightarrow \underline{\mu}(S).$$

Furthermore, it stops after n steps, where n is the number of states of the automaton.

Proof. See Section 3.5.3.

3.4.2 A Numerical Example

A numerical example is now presented which illustrates the application of Algorithm 3.4.1 to an uncertain DFSA. Consider the DFSA in Figure 3.1, where the admissible values for each event cost are presented. Note that it contains both types of uncertainty. It has not only uncertainty in the event costs but also the transitions from q_2 to q_2 and from q_3 to q_3 might not exist.

One starts by picking Δ_0. The one leading to the state transition matrix below was chosen.

$$\Pi(\Delta^0) = \begin{bmatrix} 0.3 & 0.3 & 0.2 \\ 0.3 & 0.2 & 0.4 \\ 0.4 & 0.2 & 0.2 \end{bmatrix} ; \quad \mu(\Delta^0) = \begin{bmatrix} 0 \\ 0.7143 \\ -1.0714 \end{bmatrix}$$

Since the obtained performance has $\mu_2(\Delta^0) > 0$ and $\mu_3(\Delta^0) < 0$, one now picks Δ^1 that minimizes the cost of the events leading to q_2 and maximizes the cost of the events leading to q_3. This leads to the event cost matrix and performance

---

[4] Note that this such a Δ^{k+1} can always be found since the uncertainty in each entry of the matrix $\Pi(S, \Delta)$ is independent of the uncertainty in the other entries

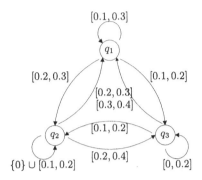

Fig. 3.1. Example of an uncertain DFSA

$$\Pi(\Delta^1) = \begin{bmatrix} 0.3 & 0.2 & 0.2 \\ 0.3 & 0 & 0.4 \\ 0.4 & 0.1 & 0.2 \end{bmatrix} ; \quad \mu(\Delta^1) = \begin{bmatrix} -0.2732 \\ 0.3825 \\ -1.3388 \end{bmatrix}$$

Since the cost of events leading to states q_1 with $\mu_1(\Delta^1) < 0$ and q_3 with $\mu_3(\Delta^1) < 0$ are at their maximum values and the cost of the events leading to q_2 with $\mu_2(\Delta^1) > 0$ are at their minimum values, the algorithm stops leading to worst-case performance

$$\underline{\mu} = \mu(\Delta^1) = \begin{bmatrix} -0.2732 & 0.3825 & -1.3388 \end{bmatrix}^T .$$

Note that the worst-case performance corresponds to a DFSA where the transition from q_2 to q_2 does not exist. The iterations of the algorithm are depicted in Figure 3.2.

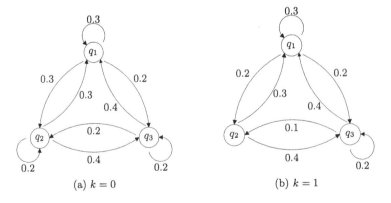

Fig. 3.2. Algorithm iterations

3.5 Proofs of Lemma 3.4.1 and Theorem 3.4.1

This section is devoted to the proofs of Lemma 3.4.1 and Theorem 3.4.1 and it can be skipped if the reader is solely interested in the application of the results in this chapter.

3.5.1 Additional Notation

Additional notation is now introduced in order to simplify the exposition to follow. Given a supervisor S and uncertainty value $\Delta \in \boldsymbol{\Delta}$, let $T_\Delta^S : \mathbf{R}^n \to \mathbf{R}^n$ be defined as

$$T_\Delta^S(\mu) \doteq \Pi(S, \Delta)\mu + X$$

Furthermore, let $T^S : \mathbf{R}^n \to \mathbf{R}^n$ be given by

$$T^S(\mu) = \min_{\Delta \in \boldsymbol{\Delta}} T_\Delta^S(\mu)$$

where the above minimum is taken entry by entry. Note that $T_\Delta^S(\cdot)$ is well-defined since, as mentioned in Section 3.3.1, the uncertainty in each entry of $\Pi(S, \Delta)$ is independent of the uncertainties in all other entries. Finally, given $x \in \mathbf{R}^n$, define the max-norm:

$$\|x\| = \max_i |x_i|.$$

Given $x, y \in \mathbf{R}^n$, it follows that $x \leq y$ if $x_i \leq y_i$ for all $i = 1, 2, \ldots, n$. It also follows that $x < y$ if $x \leq y$ and $x_i < y_i$ for some i.

Before providing the proofs of Lemma 3.4.1 and Theorem 3.4.1, a number of relevant properties of the functions $T_\Delta^S(\cdot)$ and $T^S(\cdot)$ are established as supporting lemmas.

Lemma 3.5.1 *Let S be a supervisor and $\Delta \in \boldsymbol{\Delta}$ be given, then T_Δ^S is a contraction.*

Proof. Recall that there exists a $\theta \in (0, 1)$ such that $\sum_{j=1}^{n} \pi_{ij}(S, \Delta) \leq 1 - \theta$ for all control policies. Now, let $x, y \in \mathbf{R}^n$ be two vectors, then the ith coordinate of $T_\Delta^S(x) - T_\Delta^S(y)$ satisfies the following inequality

$$\left|(T_\Delta^S(x) - T_\Delta^S(y))_i\right| = |[\Pi(S, \Delta)(x - y)]_i| \leq (1 - \theta) \|x - y\|$$

Hence

$$\left\|T_\Delta^S(x) - T_\Delta^S(y)\right\| \leq (1 - \theta) \|x - y\|.$$

This proof is completed by noting that $0 < \theta < 1$.

Lemma 3.5.2 *Let S and S' be two supervisors and let $\Delta \in \boldsymbol{\Delta}$ be given. If $T_\Delta^S(\mu(S', \Delta)) \geq \mu(S', \Delta)$ then*

$$\mu(S, \Delta) \geq \mu(S', \Delta).$$

Proof. Note that, since all entries of $\Pi(S, \Delta)$ are non-negative then

$$x \geq y \Rightarrow T^S_\Delta(x) = \Pi(S, \Delta)x + X \geq \Pi(S, \Delta)y + X = T^S_\Delta(y).$$

Therefore, it follows that

$$(T^S_\Delta)^2(\mu(S', \Delta)) \doteq T^S_\Delta\left[T^S_\Delta(\mu(S', \Delta))\right] \geq T^S_\Delta(\mu(S', \Delta)) \geq \mu(S', \Delta)$$

Repeating the above reasoning, for any positive integer k,

$$(T^S_\Delta)^k(\mu(S', \Delta)) \geq \mu(S', \Delta)$$

Now by Lemma 3.5.1 and using the Contraction Mapping Theorem [8], it follows that

$$\mu(S, \Delta) = \lim_{k \to \infty}(T^S_\Delta)^k(\mu(S', \Delta)) \geq \mu(S', \Delta).$$

Corollary 3.5.1 *Let S and S' be two supervisors. Then, $T^S_\Delta(\mu(S', \Delta)) > \mu(S', \Delta)$ implies that*

$$\mu(S, \Delta) > \mu(S', \Delta).$$

Proof. By Lemma 3.5.2 it is known that $\mu(S, \Delta) \geq \mu(S', \Delta)$ If it is assumed that $\mu(S, \Delta) = \mu(S', \Delta)$, then

$$T^S_\Delta(\mu(S', \Delta)) = T^S_\Delta(\mu(S, \Delta)) = \mu(S, \Delta) = \mu(S', \Delta)$$

and a contradiction is reached.

Corollary 3.5.2 *Let S be a controllable supervisor. Then, the operator T^S is a contraction.*

Proof. Let $x, y \in \mathbf{R}^n$ be two vectors and let Δ_y be the control policy satisfying $T^S(y) = T^S_{\Delta_y}(y)$. Then,

$$\begin{aligned}
T^S(x) - T^S(y) &= \min_{\Delta \in \Delta} T^S_\Delta(x) - \min_{\Delta \in \Delta} T^S_\Delta(y) \\
&= \min_{\Delta \in \Delta} T^S_\Delta(x) - T^S_{\Delta_y}(y) \leq T^S_{\Delta_y}(x) - T^S_{\Delta_y}(y).
\end{aligned}$$

Hence,

$$(T^S(x) - T^S(y))_i \leq [\Pi(S, \Delta_y)(x - y)]_i \leq |[\Pi(S, \Delta_y)(x - y)]_i| \leq (1 - \theta)\|x - y\|$$

Exchanging the roles of x and y, it follows that

$$(T^S(y) - T^S(x))_i \leq (1 - \theta)\|x - y\|.$$

Hence,

$$\left|(T^S(x) - T^S(y))_i\right| \leq (1 - \theta)\|x - y\|$$

and

$$\left\|(T^S(x) - T^S(y))\right\| \leq (1 - \theta)\|x - y\|$$

The proof is completed by noting that $0 < \theta < 1$.

Lemma 3.5.3 *Let S and S' be two supervisors, if $T^S(\underline{\mu}(S')) \geq \underline{\mu}(S')$ then $\underline{\mu}(S) \geq \underline{\mu}(S')$. In addition, if $T^S(\underline{\mu}(S')) > \underline{\mu}(S')$ then $\underline{\mu}(S) > \underline{\mu}(S')$.*

Proof. Similar to the proofs of Lemma 3.5.2 and Corollary 3.5.1.

Having these preliminary results, on can now proceed with the proofs of Lemma 3.4.1 and Theorem 3.4.1.

3.5.2 Proof of Lemma 3.4.1

First, note that, by Corollary 3.5.2, T^S is a contraction. Hence, there exists a $\underline{\mu}(S)$ such that

$$\underline{\mu}(S) = T^S(\underline{\mu}(S)) = \min_{\Delta \in \mathbf{\Delta}} T_\Delta^S(\underline{\mu}(S)).$$

Since $T_\Delta^S(\underline{\mu}(S))$ depends continuously on Δ and the minimization is done over the compact set $\mathbf{\Delta}$, then there exists a $\Delta^* \in \mathbf{\Delta}$ such that

$$\underline{\mu}(S) = T_{\Delta^*}^S(\underline{\mu}(S))$$

and, therefore,

$$\underline{\mu}(S) = \mu(S, \Delta^*).$$

Furthermore, by Lemma 3.5.3, T^S is a monotone operator, i.e.,

$$x \geq y \Rightarrow T^S(x) \geq T^S(y).$$

Moreover, given any $\Delta \in \mathbf{\Delta}$, Lemma 3.5.2 indicates that

$$x \geq y \Rightarrow T_\Delta^S(x) \geq T_\Delta^S(y).$$

Now, given any Δ and the associated performance $\mu(S, \Delta)$, the definition of T^S implies that

$$T^S(\mu(S, \Delta)) \leq T_\Delta^S(\mu(S, \Delta)) = \mu(S, \Delta).$$

Hence

$$\underline{\mu}(S) = \lim_{k \to \infty} (T^S)^k(\mu(S, \Delta)) \leq \mu(S, \Delta)$$

Therefore, $\underline{\mu}(S) = \mu(S, \Delta^*)$ and the proof is complete.

3.5.3 Proof of Theorem 3.4.1

Since π_{ij} multiplies μ_j in the computation of $T(\mu)$, it follows that Δ^{k+1} in Step 1 of Algorithm 3.4.1 must satisfy the condition

$$T_{\Delta^{k+1}}^S(\mu(S, \Delta^k)) = \min_{\Delta \in \mathbf{\Delta}} T_\Delta^S(\mu(S, \Delta^k)) \leq T_{\Delta^k}^S(\mu(S, \Delta^k)) = \mu(S, \Delta^k)$$

Hence, by Lemma 3.5.2,

$$\mu(S, \Delta^{k+1}) \le \mu(S, \Delta^k).$$

Moreover, the algorithm stops only if

$$\mu(S, \Delta^{k+1}) = \mu(S, \Delta^k).$$

Therefore, the stopping rule is equivalent to having

$$\mu(S, \Delta^{k+1}) = T_{\Delta^{k+1}}^S(\mu(S, \Delta^{k+1})) = T_{\Delta^{k+1}}^S(\mu(S, \Delta^k)) = \min_{\Delta \in \boldsymbol{\Delta}} T_\Delta^S(\mu(S, \Delta^k)).$$

Hence,

$$\mu(S, \Delta^k) = \min_{\Delta \in \boldsymbol{\Delta}} T_\Delta^S(\mu(S, \Delta^k)) = \underline{\mu}(S).$$

To prove convergence in n steps, note that $\mu(S, \Delta^{k+1}) \ne \mu(S, \Delta^k)$ if and only if there exists a j such that the sign of $\mu_j(S, \Delta^{k+1})$ is different from the sign of $\mu_j(S, \Delta^k)$. Moreover,

$$\mu(S, \Delta^{k+1}) \le \mu(S, \Delta^k).$$

The above monotonicity property implies that each entry of $\mu_j(S, \Delta^k)$ changes sign at most one time and, since $\mu(S, \Delta^k)$ is a vector of dimension n, the algorithm will stop after n steps.

3.6 Optimal Robust Supervisor

This section is dedicated to the development of an algorithm for robust optimal control design. This algorithm combines the worst-case performance analysis algorithm 3.4.1 with some of the ideas on optimal supervisor design presented in [5] in order to obtain a sequence of supervisors which converges to the most permissive robust optimal supervisor in a finite number of steps.

Algorithm 3.6.1 *Optimal robust supervisor design*

Step 0. Let S^0 be a controllable supervisor and let $k = 0$.

Step 1. Given S^k, determine $\underline{\mu}(S^k)$ using Algorithm 3.4.1.

Step 2. Determine S^{k+1} by disabling all the controllable events leading to states with $\underline{\mu}_j(S^k) < 0$ and enabling all events leading to states with $\underline{\mu}_j(S^k) \ge 0$.

Step 3. If $S^{k+1} = S^k$ then set $S^* = S^k$, $\underline{\mu}(S^*) = \underline{\mu}(S^k)$ and stop. Else return to Step 1 with $k \leftarrow (k + 1)$.

Next, it is shown that the above algorithms converges to the robust optimal supervisor in a finite number of steps. Furthermore, the supervisor obtained is the maximally permissive one among all optimal supervisors.

Theorem 3.6.1 *The control policy S^* obtained by Algorithm 3.6.1 is an optimal control policy over the uncertainty range and the robust performance of the closed-loop system is given by $\underline{\mu}(S^*)$. Moreover, S^* is the maximally permissive controller among all robust optimal controllers. Furthermore, the algorithm terminates in at most n steps, where n is the number of states of the (deterministic) automaton of the plant model.*

Proof. The proof is provided in the next section.

3.7 Proof of Theorem 3.6.1

This section presents the proof of Theorem 3.6.1 and it can be skipped by the readers who are solely interested in the application of the results in this chapter.

We start by introducing an additional function: Let $T : \mathbf{R}^n \to \mathbf{R}^n$ be defined as

$$T(\mu) \doteq \max_S T^S(\mu).$$

Some relevant properties of $T(\cdot)$ are established in the following two lemmas. The proofs are omitted since they are similar to the proofs of the lemmas in Section 3.5.

Lemma 3.7.1 *The transformation T is a contraction.*

Lemma 3.7.2 *There exists a S^* such that*

$$\underline{\mu}^* = \underline{\mu}(S^*) = T(\underline{\mu}(S^*)).$$

Furthermore, for all supervisor S, $\underline{\mu}^* \geq \underline{\mu}(S)$.

3.7.1 Proof of Theorem 3.6.1

The proof is started by noting that S^{k+1} in Step 1 of Algorithm 3.6.1 is chosen in such a way as to maximize π_{ij} when $\mu_j(S^k) \geq 0$ and minimize π_{ij} when $\mu_j(S^k) < 0$. Then, it follows that

$$T^{S^{k+1}}(\underline{\mu}(S^k)) = \max_S T^S(\underline{\mu}(S^k)) \geq T^{S^k}(\underline{\mu}(S^k)) = \underline{\mu}(S^k)$$

Hence, by Lemma 3.5.3,

$$\underline{\mu}(S^{k+1}) \geq \underline{\mu}(S^k).$$

Moreover, the algorithm stops only if

$$\underline{\mu}(S^{k+1}) = \underline{\mu}(S^k)$$

yielding

$$\underline{\mu}(S^{k+1}) = T^{S^{k+1}}(\underline{\mu}(S^{k+1})) = T^{S^{k+1}}(\underline{\mu}(S^k)) = \max_S T^S(\underline{\mu}(S^k)).$$

Therefore,

$$\underline{\mu}(S^k) = \max_S T^S(\underline{\mu}(S^k)) = \underline{\mu}(S^*)$$

where S^* is the optimal controller. To prove convergence in n steps, note that $\underline{\mu}(S^{k+1}) \neq \underline{\mu}(S^k)$ if and only if there exists a j such that the sign of $\underline{\mu}_j(S^{k+1})$ is different from the sign of $\underline{\mu}_j(S^k)$. Because of the fact that

$$\underline{\mu}(S^{k+1}) \geq \underline{\mu}(S^k),$$

this monotonicity property implies that each entry of $\underline{\mu}_j(S^k)$ changes sign at most one time and, since $\underline{\mu}_j(S^k)$ is a vector of dimension n, the algorithm will stop after at most n steps.

To prove that the controller S^* obtained is the most permissive one among all the optimal controllers, let $\underline{\mu}^*$ be the optimal performance and let $\mathcal{S}^*$ be the set of all optimal supervisors. Then,

$$\mathcal{S}^* = \{S : T(\underline{\mu}^*) = T^S(\underline{\mu}^*) = \underline{\mu}^*\}.$$

Given this, all supervisors in $\mathcal{S}^*$ only differ in the enabling or disabling or events leading to states q_j with optimal performance $\underline{\mu}_j^* = 0$. Since the supervisor $S^* \in \mathcal{S}^*$ obtained by the algorithm enables all events leading to states q_i with $\underline{\mu}_i^* \geq 0$, then it is the most permissive one among all optimal supervisors in $\mathcal{S}^*$.

3.8 An Application Example

As an example of the application of the robust control design algorithm proposed in this chapter, consider the design of a discrete-event robust optimal supervisor for the 3-processor decoding problem. Figure 3.3 depicts the arrangement of three processors, p_1, p_2 and p_3. With some probability, each processor may receive encoded messages for decoding. There is a server in between any 2 processors, so the total number of servers is three. On each server there is a decoding key and, at any given time, the key can only be assessed by one of the processors adjacent to the server through the link connecting the processor and the server.

In order to decode the message, the processor has to get both keys stored on the servers next to it, one at a time. After decoding, the processor will release both keys at the same time. As noted in Figure 3.3, the three processors are marked as p_1, p_2 and p_3. The servers are marked as s_1, s_2 and s_3. The keys are marked as k_1, k_2 and k_3; and L represents the link between a processor and a server with L_{ij} denoting the link connecting processor p_i and server

Table 3.1. Intervals for Event Costs and State Weights.

	p11 (1)	p13 (2)	p21 (3)	p22 (4)	p32 (5)	p33 (6)	f1 (7)	f2 (8)	f3 (9)	x
1	[0.13, 0.162]	[0.08, 0.13]	[0.14, 0.15]	[0.12, 0.158]	[0.15, 0.17]	[0.28, 0.3]	0	0	0	0
2	0	[0.09, 0.15]	0	[0.23, 0.25]	[0.2, 0.25]	[0.24, 0.3]	0	0	0	0
3	[0.18, 0.25]	0	[0.17, 0.25]	[0.2, 0.24]	[0.18, 0.24]	0	0	0	0	0
4	[0.24, 0.25]	[0.095, 0.18]	[0.23, 0.25]	0	0	[0.29, 0.3]	0	0	0	0
5	0	[0.1, 0.16]	0	[0.23, 0.25]	[0.24, 0.28]	[0.288, 0.3]	0	0	0	0
6	[0.19, 0.3]	0	[0.16, 0.2]	[0.18, 0.25]	[0.19, 0.2]	0	0	0	0	0
7	[0.23, 0.25]	[0.095, 0.1]	[0.24, 0.25]	0	0	[0.28, 0.3]	0	0	0	0
8	0	0	0	[0.3, 0.35]	[0.33, 0.35]	0	[0.16, 0.2]	0	0	0.01
9	0	0	0	0	0	0	[0.45, 0.8]	0	0	0.01
10	0	0	0	0	0	0	[0.45, 0.7]	0	0	0.01
11	0	0.1	0	0	0	[0.27, 0.3]	0	[0.29, 0.3]	0	0.02
12	0	0	0	0	0	0	0	[0.47, 0.5]	0	0.02
13	0	0	0	0	0	0	0	[0.48, 0.6]	0	0.02
14	[0.29, 0.3]	0	[0.27, 0.3]	0	0	0	0	0	[0.29, 0.3]	0.04
15	0	0	0	0	0	0	0	0	[0.389, 0.6]	0.04
16	0	0	0	0	0	0	0	0	[0.36, 0.56]	0.04
17	0	0	0	[0.03, 0.35]	[0.4, 0.65]	0	0	0	0	0
18	0	[0.1, 0.2]	0	0	0	[0.28, 0.45]	0	0	0	0
19	0	[0.1, 0.68]	0	0	0	[0.01, 0.3]	0	0	0	0
20	[0.397, 0.45]	0	[0.37, 0.46]	0	0	0	0	0	0	0
21	[0.4, 0.5]	0	[0.38, 0.4]	0	0	0	0	0	0	0
22	0	0	0	[0.38, 0.46]	[0.37, 0.48]	0	0	0	0	0
23	[0.38, 0.4]	0	[0.39, 0.54]	0	0	0	0	0	0	0
24	0	0	0	[0.39, 0.57]	[0.39, 0.4]	0	0	0	0	0
25	0	[0.1, 0.15]	0	0	0	[0.295, 0.33]	0	0	0	0
26	0	0	0	0	0	0	0	0	0	−1
27	0	0	0	0	0	0	0	0	0	−1

Table 3.2. Nominal Event Costs.

	$p11$ (1)	$p13$ (2)	$p21$ (3)	$p22$ (4)	$p32$ (5)	$p33$ (6)	$f1$ (7)	$f2$ (8)	$f3$ (9)
1	0.15	0.1	0.15	0.15	0.15	0.3	0	0	0
2	0	0.1	0	0.25	0.25	0.3	0	0	0
3	0.2	0	0.2	0.2	0.2	0	0	0	0
4	0.25	0.1	0.25	0	0	0.3	0	0	0
5	0	0.1	0	0.25	0.25	0.3	0	0	0
6	0.2	0	0.2	0.2	0.2	0	0	0	0
7	0.25	0.1	0.25	0	0	0.3	0	0	0
8	0	0	0	0.35	0.35	0	0.2	0	0
9	0	0	0	0	0	0	0.5	0	0
10	0	0	0	0	0	0	0.5	0	0
11	0	0.1	0	0	0	0.3	0	0.3	0
12	0	0	0	0	0	0	0	0.5	0
13	0	0	0	0	0	0	0	0.5	0
14	0.3	0	0.3	0	0	0	0	0	0.3
15	0	0	0	0	0	0	0	0	0.4
16	0	0	0	0	0	0	0	0	0.4
17	0	0	0	0.03	0.65	0	0	0	0
18	0	0.1	0	0	0	0.3	0	0	0
19	0	0.65	0	0	0	0.015	0	0	0
20	0.4	0	0.4	0	0	0	0	0	0
21	0.4	0	0.4	0	0	0	0	0	0
22	0	0	0	0.4	0.4	0	0	0	0
23	0.4	0	0.4	0	0	0	0	0	0
24	0	0	0	0.4	0.4	0	0	0	0
25	0	0.1	0	0	0	0.3	0	0	0
26	0	0	0	0	0	0	0	0	0
27	0	0	0	0	0	0	0	0	0

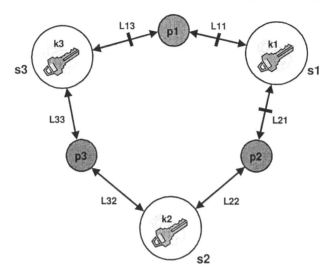

Fig. 3.3. Arrangement of connections in the 3-processor decoding problem

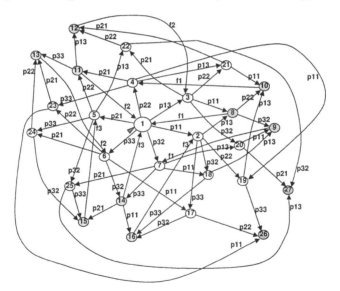

Fig. 3.4. Unsupervised plant automaton for the 3-processor decoding problem

s_j. There is a switch on the links L_{11}, L_{13} and L_{21}, so the connection can be disabled if necessary.

Figure 3.4 shows the unsupervised (i.e., open loop) plant automaton model, where state 1 is the initial state; the states 26 and 27, shaded in red, are bad marked states representing deadlocks; the states, 8, 9, 10, 11, 12, 13, 14, 15 and 16, shaded in gray, are good marked states; and the remaining unshaded states are unmarked. The event p_{ij} indicates that processor p_i has accessed key k_j. Event f_i indicates that processor p_i has finished decoding and has released both keys in its possession. The events f_i are uncontrollable. In addition, given the available switches, one can disable p_{11}, p_{13} and p_{21}, but not p_{22}, p_{32} and p_{33}. Processor p_3 normally receives the most important messages, and p_1 receives the least important messages. The good states are those states representing the successful decoding of each processor. The worst states are those states (state 26 and state 27) where each processor is holding one key and, hence, no one releases its key. Those two states have the most negative characteristic function values.

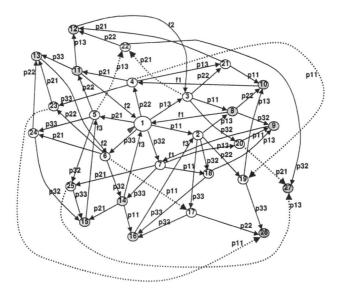

Fig. 3.5. Optimal control for nominal event costs

For the analysis of robust optimal control, nominal values and uncertainty bounds are specified for each event cost. The values used are provided in Tables 3.1 and 3.2. The supervisor is synthesized using the optimal control policy obtained by applying the results in [5]. Figure 3.5 shows optimal supervisor designed for nominal values of event cost, where the disabled events

are represented by dotted lines; controllable events, causing state transitions
$3 \rightarrow 32$, $4 \rightarrow 19$, $5 \rightarrow 22$, $6 \rightarrow 17$, $20 \rightarrow 27$, $23 \rightarrow 26$, and $25 \rightarrow 27$, are
disabled under the control. Further analysis indicates that the performance of
this controller degrades significantly for other uncertainties values.

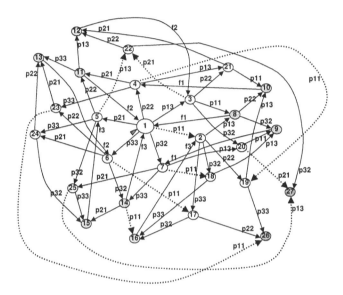

Fig. 3.6. Robust Optimal control for uncertain event costs

Let us apply the algorithms proposed in this chapter and determine the
controller that achieves the best worst-case performance over the specified
uncertainty range. This controller may not yield optimal performance for the
given uncertainty, but it will have best performance in the worst-case scenario.
Figure 3.6 shows a finite-state machine representation of the robust optimal
controller where, again, the disabled events are represented by dotted lines;
controllable events, causing state transitions $3 \rightarrow 32$, $4 \rightarrow 19$, $5 \rightarrow 22$, $6 \rightarrow$
17, $14 \rightarrow 16$, $20 \rightarrow 27$, $23 \rightarrow 26$, and $25 \rightarrow 27$, are disabled under the control.
This controller disables one additional arc from $14 \rightarrow 16$ and hence is different
form the optimal controller for nominal event costs. In general, the optimal
controller for nominal performance may not be the robust optimal controller
under uncertain parameters of the language measure.

3.9 Summary and Conclusions

This chapter presents the theory of a state-based robust optimal control policy of regular languages for finite state automata that may have already been

subjected to constraints such as control specifications [11]. The synthesis procedure is quantitative and relies on a signed real measure of formal languages, which is based on a specified state transition cost matrix and a characteristic vector [16]. The objective is to maximize the worst-case performance vector that is obtained over the event cost uncertainty range without any further constraints. The robust optimal control policy maximizes the performance vector by selectively disabling controllable events that may lead to bad marked states and simultaneously ensuring that the remaining controllable events are kept enabled. The worst-case performance is guaranteed within the specified event cost uncertainty bounds. The control policy induced by the updated state transition cost matrix yields maximal performance and is unique in the sense that the controlled language is most permissive (i.e., least restrictive) among all controller(s) having the robust optimal performance. The computational complexity of the robust optimal control synthesis procedure is polynomial in the number of automaton states.

Future areas of research in robust optimal control include: (i) incorporation of the cost of disabling controllable events [4] and (ii) robustness of the control policy relative to other uncertainties in the plant model including loss of controllability and observability.

References

1. Y. Brave and M. Heyman, *On optimal attraction of discrete-event processes*, Information Science **67** (1993), no. 3, 245–276.
2. J. E. R. Cury and B. H. Krogh, *Robustness of supervisors for discrete-event systems*, IEEE Transactions on Automatic Control **44** (1999), no. 2, 376–379.
3. J. Fu, C.M. Lagoa, and A. Ray, *Robust optimal control of regular languages with event cost uncertainties*, Proceedings of IEEE Conference on Decision and Control, December 2003, pp. 3209–3214.
4. J. Fu, A. Ray, and C.M. Lagoa, *Optimal control of regular languages with event disabling cost*, Proceedings of American Control Conference, Denver, Colorado, June 2003, pp. 1691–1695.
5. J. Fu, A. Ray, and C.M. Lagoa, *Unconstrained optimal control of regular languages*, Automatica **40** (2004), no. 4, 639–648.
6. R. Kumar and V. Garg, *Optimal supervisory control of discrete event dynamical systems*, SIAM J. Control and Optimization **33** (1995), no. 2, 419–439.
7. F. Lin, *Robust and adaptive supervisory control of discrete event systems*, IEEE Transactions on Automatic Control **38** (1993), no. 12, 1848–1852.
8. A.W. Naylor and G.R. Sell, *Linear operator theory in engineering and science*, Springer-Verlag, 1982.
9. S.-J. Park and J.-T. Lim, *Robust and nonblocking supervisory control of nondeterministic discrete event systems using trajectory models*, IEEE Transactions on Automatic Control **47** (2002), no. 4, 655–658.
10. K. Passino and P. Antsaklis, *On the optimal control of discrete event systems*, Proceedings of 28th IEEE Decision and Control Conference, Tampa, Florida (1989), 2713–2718.

11. P.J. Ramadge and W.M. Wonham, *Supervisory control of a class of discrete event processes*, SIAM J. Control and Optimization **25** (1987), no. 1, 206–230.
12. A. Ray and S. Phoha, *Signed real measure of regular languages for discrete-event automata*, Int. J. Control **76** (2003), no. 18, 1800–1808.
13. R. Sengupta and S. Lafortune, *An optimal control theory for discrete event systems*, SIAM J. Control and Optimization **36** (1998), no. 2, 488–541.
14. A. Surana and A. Ray, *Signed real measure of regular languages*, Demonstratio Mathematica **37** (2004), no. 2, 485–503.
15. S. Takai, *Design of robust supervisors for prefix-closed language specifications*, Proceedings of the 38th IEEE Conference on Decision and Control, 1999, pp. 1725–1730.
16. X. Wang and A. Ray, *A language measure for performance evaluation of discrete-event supervisory control systems*, Applied Mathematical Modelling **28** (2004), no. 9, 817–833.

4

Advanced Topics in Supervisory Control

Ishanu Chattopadhyay[1] and Asok Ray[2]

[1] The Pennsylvania State University, PA 16802 ixc128@psu.edu
[2] The Pennsylvania State University, PA 16802 axr2@psu.edu

Summary. The signed real measure of regular languages, introduced in Chapter 1, has been the driving force for quantitative analysis and synthesis of discrete-event supervisory (DES) control systems dealing with finite state automata (equivalently, regular languages). However, this approach relies on memoryless state-based tools for supervisory control synthesis and may become inadequate if the transitions in the plant dynamics cannot be captured by finitely many states. From this perspective, the measure of regular languages needs to be extended to that of non-regular languages, such as Petri nets or other higher level languages in the Chomsky hierarchy [9]. The development of measures for non-regular languages is a topic of future research that has not apparently been reported in open literature. This chapter introduces two research topics in the field of language-measure-based supervisory control. One topic is complex measure of non-regular languages, dealing with *linear context free grammars* (*LCFG*). The proposed complex measure reduces to the signed real measure [16] [12] [15], as presented in Chapter 1, if the *LCFG* is degenerated to a regular grammar. The other topic is modification of the (regular) language measure for supervisory control under partial observation. This chapter shows how to generalize the analysis to situations where some of the events may not be observable at the supervisory level.

Key words: Discrete Event Supervisory Control, Language Measure, Non-regular Languages, Partial Observation

4.1 Introduction

Finite state automata (FSA) (equivalently, regular languages) have been widely used to model and synthesize supervisory control laws for discrete-event plants [11] [8] [1]. The mathematical simplicity of regular languages makes the control synthesis computationally efficient. According to the paradigm of supervisory control, a finite sate automaton (e.g., the discrete event model of a physical plant) is a language generator whose behavior is constrained by a supervisor to meet a control specification. For a given application, the (controlled) sublanguage of the plant behavior could be different under different supervisors that satisfy their own respective specifications. Such a partially ordered set of sublanguages requires a quantitative measure for total ordering of their respective performance. To address this issue, a signed real measure has been formulated and reported in literature [16] [12] [15] to provide a mathematical framework for quantitative comparison of sublanguages of a regular language. This measure formalizes a procedure for design of discrete event supervisory (DES) controllers for finite state automaton plants, as an alternative to the procedure of Ramadge and Wonham [11]. Chapters 2 and 3 have reported quantitative analysis and synthesis of optimal and robust control laws for finite state automata based on the language measure [6] [5] [4]. The approach is state-based and the language measure parameters are identified from experiments on the physical process or from simulation on a deterministic finite state automaton (DFSA) model of the plant [17]. However, using memoryless state-based tools for supervisory control synthesis may suffer serious shortcomings if the details of transitions cannot be captured by finitely many states. This problem has been partially circumvented by Petri nets that can accommodate certain classes of non-regular languages [10] in the Chomsky hierarchy [9]. There is apparently no quantitative tool for supervisory control synthesis of Petri nets compared to what are available for finite state automata [6] [5] [4]. Hence, there is a need for developing quantitative tools of supervisory control synthesis for discrete-event systems that cannot be represented by regular languages. Toward achieving this goal, the first step is to construct measure(s) of non-regular languages where the state-based approach [16] [12] may not be applicable.

It is first shown that the measure of a regular language proposed in [16] [12] [15] is equivalent to that of the regular grammar, without referring to states of the automaton. Then, it extends the signed real measure of regular languages to a complex measure for the class of non-regular languages [2], generated by *linear context free grammar* (*LCFG*) that is a subclass of *pushdown automata* (PDA) [7]. The signed real measure [16] [12] is extended to a complex measure over the real field, where the multiplication operation of complex numbers is different from the conventional one. In this case, the complex space over the real field degenerates to the union of a pair of one-dimensional real spaces instead of being isomorphic to the two-dimensional real space. The extended complex-valued language measure, formulated in this chapter, is potentially applicable to analysis and synthesis of DES control laws where the plant model could be represented by an *LCFG*.

The language measure introduced in [16] [12] has not addressed the issue of unobservable events at the supervisory level. This is a nontrivial restriction from the perspective of DES control synthesis since the model of a given physical process may have inherently unobservable transitions. Furthermore, in complex engineering systems, events may become unobservable due to failures of sensors or communication links in one or more locations.

This chapter extends the work reported in [16] [12] [15] to scenarios where the set of unobservable events is allowed to be nonempty. The central idea behind this generalization is the fact that the observed initiation or termination state of a given string may not be the true initiation or termination states due to the possibility of unobservable transitions.

The chapter is organized in eight sections including the present one. Section 4.2 introduces notations and background materials along with definitions of the key concepts. Section 4.3 summarizes the signed real measure of regular languages. Section 4.4 discusses the measure of regular grammars and shows its equivalence with that of regular languages. Section 4.5 extends the measure to linear grammars and the concept is elucidated with an example. Section 4.6 generalizes the signed real measure for regular grammars to plants (modeled as DFSAs) with possibly nonempty set of unobservable events. Section 4.7

prposes a method to quantify the effects of unobservability in terms of the language measure. The chapter is summarized and concluded in Section 4.8.

4.2 Concepts and Notations

This section introduces notations and background materials for formal languages along with definitions of the key concepts.

Definition 4.2.1 *A context free grammar (CFG) is a 4-tuple $\Gamma = (V, T, P, S)$, where V and T are mutually disjoint (i.e., $V \cap T = \emptyset$) finite sets of variables and terminals, respectively; and P is a finite set of productions by which strings are derived from the start symbol S. Each production in P is of the form $v \rightarrow \alpha$, where $v \in V$ and $\alpha \in (V \cup T)^*$, where the superscript $*$ indicates Kleene closure [7].*

Remark 4.2.1 *The language generated by a grammar Γ consists of all strings obtained from legal (i.e., permissible) productions beginning with the start symbol.*

Definition 4.2.2 *A regular grammar is a CFG (V, T, P, S) where every production in P takes exactly one of the following two alternative pairs of forms (i.e., there are either right derivations or left derivations but not both):*

$$
\begin{cases} v \rightarrow \alpha w \\ v \rightarrow \alpha \end{cases} \quad \text{or} \quad \begin{cases} v \rightarrow w\alpha \\ v \rightarrow \alpha \end{cases} \tag{4.1}
$$

where $v, w \in V$ and $\alpha \in T \cup \{\epsilon\}$; and ϵ is the empty string.

Remark 4.2.2 *The generated language for a deterministic finite state automaton (DFSA) is a regular language [7].*

Definition 4.2.3 *A context free grammar is defined to be linear if no right side of a production has more than one instance of a variable. That is, for a linear context free grammar (V, T, P, S), every production in P takes one of the following forms:*

$$
v \rightarrow \alpha w; \quad v \rightarrow w\alpha; \quad v \rightarrow \alpha \tag{4.2}
$$

where $v, w \in V$ and $\alpha \in T \cup \{\epsilon\}$.

Remark 4.2.3 *In view of Remark 4.2.1 and Definition 4.2.3, the set of production rules P in a linear grammar $\Gamma = (V, T, P, S)$ can be modified as $\tilde{\Gamma} = (\tilde{V}, T, \tilde{P}, S)$ by augmenting the set V of variables as $\tilde{V} \equiv V \cup A$ and by updating the set P of production rules by $\tilde{P}$ that is given as:*

if $\varphi = (v \to \alpha) \in P$ is replaced by $\tilde{\varphi} = (v \to a\alpha)$ where $v \in V$, $\alpha \in T \cup \{\epsilon\}$, and $a \in A$. This is analogous to the trim operation in regular languages [1] [11].

Remark 4.2.4 *The modified grammar $\tilde{\Gamma} = (\tilde{V}, T, \tilde{P}, S)$ is a superset of the original grammar $\Gamma = (V, T, P, S)$ in the sense that it contains the generated language of $\Gamma = (V, T, P, S)$ and, in addition, has productions of the type $v \to aw$. The production $A \to \epsilon$ is added to P in each step of the modification; and $v \to \epsilon$ must exist $\forall v \in V$.*

Remark 4.2.5 *Regular grammars have only right (or only left) derivations with a single variable. In contrast, linear grammars include both right and left derivations with a single variable. This is precisely what allows the linear grammars to model a certain class of non-regular languages.*

In the sequel, the terms **state** and **variable** are used interchangeably as they convey the similar meaning in the present context; the same applies to the terms **terminals** and **events**.

4.3 Measure of Regular Languages

This section first summarizes the signed real measure of regular languages [16] [15] reported in Chapter 1. Then, the concept of the state-based language measure is reformulated in Section 4.4 in terms of regular grammars, followed by the construction of measure.

4.3.1 State Based Language Measure

Let $G_i \equiv \langle Q, \Sigma, \delta, q_i, Q_m \rangle$ be a trim (i.e., accessible and co-accessible) finite-state automaton model that represents the discrete-event dynamics of a physical plant, where $Q = \{q_k : k \in \mathcal{I}_Q\}$ is the set of states and $\mathcal{I}_Q \equiv \{1, 2, \cdots, n\}$ is the index set of states; the automaton starts with the initial state q_i; the alphabet of events is $\Sigma = \{\sigma_k : k \in \mathcal{I}_\Sigma\}$ with $\Sigma \cap \mathcal{I}_Q = \emptyset$, and $\mathcal{I}_\Sigma \equiv \{1, 2, \cdots, \ell\}$ is the index set of events; $\delta : Q \times \Sigma \to Q$ is the (possibly partial) function of state transitions; and $Q_m \equiv \{q_{m_1}, q_{m_2}, \cdots, q_{m_l}\} \subseteq Q$ is the set of marked (i.e., accepted) states with $q_{m_k} = q_j$ for some $j \in \mathcal{I}_Q$.

Let Σ^* be the Kleene closure of Σ, i.e., the set of all finite-length strings made of the events belonging to Σ as well as the empty string ϵ that is viewed as the identity of the monoid Σ^* under the operation of string concatenation, i.e., $\epsilon s = s = s\epsilon$. The extension $\hat{\delta} : Q \times \Sigma^* \to Q$ is defined recursively in the usual sense [7] [9] [11].

Definition 4.3.1 *The language $L(G_i)$ generated by a DFSA G initialized at the state $q_i \in Q$ is defined as:*

$$L(G_i) = \{s \in \Sigma^* \mid \hat{\delta}(q_i, s) \in Q\} \tag{4.3}$$

The language $L_m(G_i)$ marked by the DFSA G initialized at the state $q_i \in Q$ is defined as:

$$L_m(G_i) = \{s \in \Sigma^* \mid \hat{\delta}(q_i, s) \in Q_m\} \tag{4.4}$$

Definition 4.3.2 *For every $q_j \in Q$, let $L_{i,j}$ denote the set of all strings that, starting from the state q_i, terminate at the state q_j, i.e.,*

$$L_{i,j} = \{s \in \Sigma^* \mid \hat{\delta}(q_i, s) = q_j \in Q\} \tag{4.5}$$

The set Q_m of marked states is partitioned into Q_m^+ and Q_m^-, i.e., $Q_m = Q_m^+ \cup Q_m^-$ and $Q_m^+ \cap Q_m^- = \emptyset$, where Q_m^+ contains all *good* marked states that we desire to reach, and Q_m^- contains all *bad* marked states that we want to avoid, although it may not always be possible to completely avoid the *bad* states while attempting to reach the *good* states. To characterize this, each marked state is assigned a real value based on the designer's perception of its impact on the system performance.

Definition 4.3.3 *The characteristic function $\chi : Q \to [-1, 1]$ that assigns a signed real weight to state-based sublanguages $L_{i,k}$ of the DFSA G_i is defined as:*

$$\forall q_k \in Q, \quad \chi(q_k) \in \begin{cases} [-1, 0), & q_k \in Q_m^- \\ \{0\}, & q_k \notin Q_m \\ (0, 1], & q_k \in Q_m^+ \end{cases} \tag{4.6}$$

The state weighting vector, denoted by $\mathbf{X} = [\chi_1 \; \chi_2 \; \cdots \; \chi_n]^T$, where $\chi_j \equiv \chi(q_j) \; \forall j \in \mathcal{I}_Q$, is called the $\mathbf{X}$-vector. The j-th element χ_j of $\mathbf{X}$-vector is the weight assigned to the corresponding terminal state q_j.

In general, the marked language $L_m(G_i)$ consists of both good and bad event strings that, starting from the initial state q_i, lead to Q_m^+ and Q_m^- respectively. Any event string belonging to the language $L^0 = L(G_i) - L_m(G_i)$ leads to one of the non-marked states belonging to $Q - Q_m$ and L^0 does not contain any one of the good or bad strings. Based on the equivalence classes defined in the Myhill-Nerode Theorem, the regular languages $L(G_i)$ and $L_m(G_i)$ can be expressed as:

$$L(G_i) = \bigcup_{q_k \in Q} L_{i,k} \tag{4.7}$$

$$L_m(G_i) = \bigcup_{q_k \in Q_m} L_{i,k} = L_m^+ \cup L_m^- \qquad (4.8)$$

where the sublanguage $L_{i,k} \subseteq G_i$ having the initial state q_i is uniquely labelled by the terminal state $q_k, k \in \mathcal{I}_Q$ and $L_{i,j} \cap L_{i,k} = \emptyset \; \forall j \neq k$; and $L_m^+ \equiv \bigcup_{q_j \in Q_m^+} L_{i,j}$ and $L_m^- \equiv \bigcup_{q_j \in Q_m^-} L_{i,j}$ are good and bad sublanguages of $L_m(G_i)$, respectively. Then, $L^0 = \bigcup_{q_j \notin Q_m} L_{i,j}$ and $L(G_i) = L^0 \cup L_m^+ \cup L_m^-$.

A signed real measure $\mu^i : 2^{L(G_i)} \to \mathbf{R} \equiv (-\infty, +\infty)$ is constructed on the σ-algebra $2^{L(G_i)}$; interested readers are referred to [16] [15] for the details of measure-theoretic definitions and results. With the choice of this σ-algebra, every singleton set made of an event string $s \in L(G_i)$ is a measurable set. By Hahn Decomposition Theorem [14], each of these measurable sets is qualified to have a numerical value based on the above state-based decomposition of $L(G_i)$ into null, positive and negative sublanguages, denoted as L^0, L^+, and L^-, respectively.

In the following definition, each event is assigned a state-dependent cost that is conceptually similar to the conditional transition probability.

Definition 4.3.4 *The event cost of the DFSA G_i is defined as a (possibly partial) function $\tilde{\pi} : \Sigma^* \times Q \to [0, 1]$ such that $\forall q_i \in Q, \forall \sigma_j \in \Sigma, \forall s \in \Sigma^*$,*

$$\tilde{\pi}[\sigma_j, q_i] = 0 \text{ if } \delta(q_i, \sigma_j) \text{ is undefined}; \quad \tilde{\pi}[\epsilon, q_i] = 1;$$

$$\tilde{\pi}[\sigma_j, q_i] \equiv \tilde{\pi}_{ij} \in [0, 1); \quad \sum_{j \in \mathcal{I}_\Sigma} \tilde{\pi}_{ij} < 1; \qquad (4.9)$$

$$\tilde{\pi}[\sigma_j s, q_i] = \tilde{\pi}[\sigma_j, q_i] \, \tilde{\pi}[s, \delta(q_i, \sigma_j)].$$

Consequently, the $n \times \ell$ event cost matrix is defined as:

$$\widetilde{\Pi} = \begin{bmatrix} \tilde{\pi}_{11} & \tilde{\pi}_{12} & \cdots & \tilde{\pi}_{1\ell} \\ \tilde{\pi}_{21} & \tilde{\pi}_{22} & \cdots & \tilde{\pi}_{2\ell} \\ \vdots & \vdots & \ddots & \vdots \\ \tilde{\pi}_{n1} & \tilde{\pi}_{n2} & \cdots & \tilde{\pi}_{n\ell} \end{bmatrix} \qquad (4.10)$$

Definition 4.3.5 *The state transition cost, $\pi : Q \times Q \to [0, 1)$, of the DFSA G_i is defined as follows:*

$$\forall q_i, q_j \in Q, \pi_{ij} = \begin{cases} \sum_{\sigma \in \Sigma} \tilde{\pi}[\sigma, q_i], & \text{if } \delta(q_i, \sigma) = q_j \\ 0 & \text{if } \{\delta(q_i, \sigma) = q_j\} = \emptyset. \end{cases} \qquad (4.11)$$

Consequently, the $n \times n$ state transition cost matrix is defined as

$$\Pi = \begin{bmatrix} \pi_{11} & \pi_{12} & \cdots & \pi_{1n} \\ \pi_{21} & \pi_{22} & \cdots & \pi_{2n} \\ \vdots & \vdots & \ddots & \vdots \\ \pi_{n1} & \pi_{n2} & \cdots & \pi_{nn} \end{bmatrix} \qquad (4.12)$$

Definition 4.3.6 *The signed real measure μ^i of every singleton string set $\{\omega\} \in 2^{L(G_i)}$ where $\Omega \in L_{i,j}$ is defined as $\mu(\{\omega\}) \equiv \tilde{\pi}(\omega, q_i)\chi(q_j)$. It follows that the signed real measure of the sublanguage $L_{i,j} \subseteq L(G_i)$ is defined as*

$$\mu^i(L_{i,j}) = \left(\sum_{\omega \in L_{i,j}} \tilde{\pi}[\omega, q_i] \right) \chi_j \tag{4.13}$$

Therefore, the signed real measure of the language of a DFSA G_i initialized at $q_i \in Q$, is defined as

$$\mu_i \equiv \mu^i(L(G_i)) = \sum_{q \in Q} \mu^i(L_{i,j}) \tag{4.14}$$

The language measure vector, denoted as $\boldsymbol{\mu} = [\mu_1 \ \mu_2 \ \cdots \ \mu_n]^T$, is called the μ-vector.

Definition 4.3.7 *Let $L_i \equiv L(G_i), i \in \mathcal{I}_Q\}$, denote the regular expression representing the language of an n-state DFSA $G_i = \langle Q, \Sigma, \delta, q_i, Q_m \rangle$, where q_i is the initial state.*

Definition 4.3.8 *Let σ_j^k denote the set of event(s) $\sigma \in \Sigma$ that is defined on the state q_j and leads to the state q_k, where $j, k \in \mathcal{I}_Q$, i.e., $\delta(q_j, \sigma) = q_k, \forall \sigma \in \sigma_j^k \subseteq \Sigma$.*

Then, given a DFSA $G_i = \langle Q, \Sigma, \delta, q_i, Q_m \rangle$ with $|Q| = n$, the procedure to obtain the system equation by a set of regular expressions L_i of the language $L(G_i), i \in \mathcal{I}_Q\}$ follows.

$$\forall q_i \in Q, \qquad L_i = \sum_{j \in \mathcal{I}_Q} R_{i,j} + \mathcal{E}_i, \qquad i \in \mathcal{I}_Q\} \tag{4.15}$$

where $\forall i$, $R_{i,j}$ and $\mathcal{E}_i$ are defined as

1. If $\exists \ \sigma \in \Sigma$, such that $\delta(q_i, \sigma) = q_j \in Q, j \in \mathcal{I}_Q\}$, then $R_{i,j} = \sigma_i^j L_j$, otherwise, $R_{i,j} = \emptyset$.
2. If $q_i \in Q_m$, $\mathcal{E}_i = \epsilon$, otherwise, $\mathcal{E}_i = \emptyset$.

The set of symbolic equations may be written as

$$L_i = \sum_j \sigma_i^j L_j + \mathcal{E}_i \tag{4.16}$$

Lemma 4.3.1 *(Arden's Rule) Let u, v be two known regular expressions and r be an unknown regular expression that satisfies the following algebraic identity:*

$$r = ur + v \tag{4.17}$$

Then, the following relations are true:

(1) $r = u^*v$ is a solution to Equation (4.17).
(2) If $\epsilon \notin u$, then $r = u^*v$ is the unique solution to Equation (4.17).

Proof. The proof of Lemma 4.3.1 is given in [3] [12].

Instead of obtaining regular expressions, the language measure can be directly computed by transforming this set of equations into a system of linear equations based on the following result.

Theorem 4.3.1 *The language measure of the symbolic Equation 4.16 is given by*

$$\mu_i = \sum_j \pi_{ij}\mu_j + \chi_i \tag{4.18}$$

Proof. Following Equation (4.15) and Definition 6.2.3:

$$\forall i \in \mathcal{I}_Q, \qquad \mu(\mathcal{E}_i) = \begin{cases} \chi_i & \text{if } \mathcal{E}_i = \epsilon \\ 0 & \text{otherwise} \end{cases} \tag{4.19}$$

Therefore, each element of the vector $\mathbf{X} = [\chi_1\ \chi_2\ \cdots\ \chi_n]^T$ is the forcing function in Equations (4.16) and (4.18). Starting from the state q_i, the measure of the language $L_i \equiv L(G_i)$ (see Definition 4.3.7)

$$\begin{aligned}
\mu_i \equiv \mu^i(L_i) &= \mu^i\left(\sum_j \sigma_i^j L_j + \mathcal{E}_i\right) \\
&= \mu^i\left(\sum_j \sigma_i^j L_j\right) + \mu^i(\mathcal{E}_i) \\
&= \sum_j \mu^i(\sigma_i^j L_j) + \mu^i(\mathcal{E}_i) \\
&= \sum_j \pi(\sigma_i^j)\mu^j(L_j) + \mu^i(\mathcal{E}_i) \\
&= \sum_j \pi_{ij}\mu_j + \chi_i
\end{aligned}$$

The second equality in the above derivation follows from the fact that $\mathcal{E}_i \cap \sigma_i^j L_j = \emptyset$. It is also true that

$$\forall j \neq k, \qquad \sigma_i^j L_j \cap \sigma_i^k L_k = \emptyset$$

since each string in $\sigma_i^j L_j$ starts with an event in σ_i^j while each string in $\sigma_i^k L_k$ starts from an event in σ_i^k and $\sigma_i^j \cap \sigma_i^k = \emptyset$, as G_i is a DFSA. The remaining steps follow directly from the definitions of the transition cost function and the construction of the measure. A more complete justification is given in in [15].

In vector form, Equation (4.18) becomes

$$\mu = \Pi\mu + X \tag{4.20}$$

whose solution is given by

$$\mu = (I - \Pi)^{-1}X \tag{4.21}$$

where the inverse in Equation (6.10) exists because Π is a contraction operator [15].

Lemma 4.3.2 *The expression for language measure in Equation (6.10) can be restructured into the following form.*

$$\begin{bmatrix} \mu^1(L_{1,1}) & \mu^1(L_{1,2}) & \cdots & \mu^1(L_{1,n}) \\ \mu^2(L_{2,1}) & \mu^2(L_{2,2}) & \cdots & \mu^2(L_{2,n}) \\ \vdots & \vdots & \ddots & \vdots \\ \mu^n(L_{n,1}) & \mu^n(L_{n,2}) & \cdots & \mu^n(L_{n,n}) \end{bmatrix} = [I - \Pi]^{-1} \begin{bmatrix} \chi_1 & 0 & \cdots & 0 \\ 0 & \chi_2 & \cdots & 0 \\ \vdots & \vdots & \ddots & \vdots \\ 0 & 0 & \cdots & \chi_n \end{bmatrix} \tag{4.22}$$

Proof. Equation (4.16) is restated by using Definition 4.3.2 as

$$L_{i,j} = \sum_k \sigma_i^k L_{k,j} + \mathcal{E}_{i,j} \tag{4.23}$$

where $\mathcal{E}_{i,j} \equiv \begin{cases} \epsilon & \text{if } i = j \\ \emptyset & \text{otherwise} \end{cases}$

Hence it follows:

$$\mu_{i,j} \equiv \mu^i(L_{i,j}) = \mu^i\left(\sum_k \sigma_i^k L_{k,j} + \mathcal{E}_{i,j}\right)$$

$$= \mu^i\left(\sum_k \sigma_i^k L_{k,j}\right) + \mu^i(\mathcal{E}_{i,j})$$

$$= \sum_k \mu^i(\sigma_i^k L_{k,j}) + \mu^i(\mathcal{E}_{i,j})$$

$$= \sum_k \pi_{ik}\mu^k(L_{k,j}) + \mu^i(\mathcal{E}_{i,j})$$

$$= \sum_k \pi_{ik}\mu_{k,j} + \mu^i(\mathcal{E}_{i,j})$$

and $\mu^i(\mathcal{E}_{i,j}) = \begin{cases} \chi_i & \text{if } i = j \\ 0 & \text{otherwise} \end{cases}$

The steps in the above proof are similar to those for derivation of μ_i in the proof of Theorem 4.3.1.

Lemma 4.3.3 *The measure of the marked language $L_i \equiv L_m(G_i)$ is given as*

$$\mu_i \equiv \sum_j \mu_{i,j} = \sum_j \left(\sum_{\omega \in L_{i,j}} \tilde{\pi}[\omega, q_i] \right) \chi_j$$

$$= \left\{ [I - \Pi]^{-1} X \right\}_i \tag{4.24}$$

Proof. Lemma 4.3.2 implies the following result.

$$\mu_{i,j} \equiv \mu^i(L_{i,j}) = \left(\sum_{\omega \in L_{i,j}} \tilde{\pi}[\omega, q_i] \right) \chi_j$$

$$= \left\{ [I - \Pi]^{-1} \begin{bmatrix} \chi_1 & \cdots & 0 \\ \vdots & \chi_i & \vdots \\ 0 & \cdots & \chi_n \end{bmatrix} \right\}_{i,j} \tag{4.25}$$

Language measure, under under partial observability, would require summation over the indices of initial states, in addition to summing over the indices of terminal states in Lemma 4.3.3.

Lemma 4.3.4 *Summation over the indices of initial states yields the following language measure.*

$$\sum_i \mu_{i,j} = \sum_i \left(\sum_{\omega \in L_{i,j}} \tilde{\pi}[\omega, q_i] \right) \chi_j$$

$$= \left\{ D \left([I - \Pi]^{-T} \begin{Bmatrix} 1 \\ \vdots \\ 1 \end{Bmatrix} \right) X \right\}_j \tag{4.26}$$

where the diagonalization map D is defined as

$$D : R^n \longrightarrow R^{n \times n}$$

$$\text{where } D \left(\begin{Bmatrix} \xi_1 \\ \vdots \\ \xi_n \end{Bmatrix} \right) \doteq \begin{bmatrix} \xi_1 & \cdots & 0 \\ \vdots & \ddots & \vdots \\ 0 & \cdots & \xi_n \end{bmatrix} \quad \forall \xi \in R^n \tag{4.27}$$

Proof. Equation (4.26) follows from the fact that each diagonal entry of

$$D \left([I - \Pi]^{-T} \begin{Bmatrix} 1 \\ \vdots \\ 1 \end{Bmatrix} \right) \text{ is the corresponding column sum of } [I - \Pi]^{-1}.$$

4.4 Measure of Regular Grammars

This section first introduces the concept of regular-grammar-based measures and then shows its equivalence to that of recently reported state-based measure [16] [12] [15]. In essence, the concept of the state-based language measure is reformulated in terms of regular grammars, followed by construction of the measure. While detailed proofs of the supporting theorems are given in [7], sketches of the proofs that are necessary for developing the underlying theory are presented here.

Theorem 4.4.1 *If L is a regular language, then $\exists$ a regular grammar Γ such that either $L = L(\Gamma)$ or $L = L(\Gamma) \cup \{\epsilon\}$.*

Proof. Let $G = (Q, \Sigma, \delta, q_i, A)$ be an FSA. The grammar Γ is constructed with $V = Q$ and $T = \Sigma$. The set of productions is constructed as follows:

$$\forall q_i, q_j \in Q, \ s_r \in \Sigma, \begin{cases} Add \ q_i \to s_r q_j \ if \ \delta(q_i, s_r) = q_j \\ Add \ q_i \to s_r \ if \delta(q_i, s_r) \in A \end{cases} \tag{4.28}$$

Theorem 4.4.2 *If Γ is a regular grammar, then $L(\Gamma)$ is a regular language.*

Proof. Let $\Gamma = (V, T, P, S)$ be a regular grammar. A (possibly nondeterministic) finite state automaton G is constructed such that it exactly accepts the language $L(\Gamma)$. Specifically, let $G = (V \cup \{W\}, T, \delta, S, \{W\})$, where W is the only marked state and δ is defined as follows:

$$\delta(v_i, s_r) \equiv \delta_1(v_i, s_r) \cup \delta_2(v_i, s_r) \cup \delta_3(v_i, s_r)$$

$$where \begin{cases} \delta_1(v_i, s_r) = v_j, & if \ v_i \to s_r v_j \in P \\ \delta_2(v_i, s_r) = \{W\}, & if \ v_i \to s_r \in P \\ \delta_3(v_i, s_r) = \phi, & otherwise \end{cases}$$

Remark 4.4.1 *It follows from Theorem 4.4.2 and Theorem 4.4.1 that a language L is regular iff $\exists$ a regular grammar Γ such that either $L = L(\Gamma)$ or $L = L(\Gamma) \cup \{\epsilon\}$. Therefore, $\exists$ a regular grammar for every finite state automaton that exactly generates the language of the regular grammar and vice versa.*

4.4.2 Formulation of Regular Grammar Measures

This section follows the same construction procedure as in [16] [15] [12] because there exists a one-to-one-correspondence between the state set Q of an automaton and the variable set V of the corresponding grammar. The same holds true for the alphabet set Σ and the terminal T of the regular grammar. The notion of marked states as well as that of good and bad marked states translates naturally to this framework. The variable set V can be partitioned into sets of marked variables V_m and non-marked variables $V - V_m$. The set V_m is further partitioned into good and bad marked variables as V_m^+ and V_m^-.

Definition 4.4.1 *The language $L(\Gamma_i)$ generated by a context free grammar (CFG) Γ_i initialized at state $v_i \in V$ is defined as:*

$$L(\Gamma_i) = \{s \in \Sigma^* | \exists \text{ a derivation of } s \text{ from } \Gamma_i\} \tag{4.29}$$

Definition 4.4.2 *The language $L_m(\Gamma_i)$ generated by a CFG Γ_i initialized at state $v_i \in V$ is defined as:*

$L_m(\Gamma_i) = \{s \in \Sigma^* | \exists \text{ a derivation of } s \text{ from } \Gamma_i \text{ which terminates on a marked variable }\}$

Definition 4.4.3 *For every $v_i, v_k \in V$, the set of all strings that, starting from v_i, terminate on v_k is defined as the language $L(v_i, v_k)$. That is,*

$L(v_i, v_k) = \{s \in \Sigma^* | \exists \text{ a derivation of } s \text{ from } v_i \text{ that terminates on } v_k\}$

Definition 4.4.4 *The characteristic function $\chi : V \to [-1, 1]$ is defined in exact analogy with the state based approach.*

$$\forall \tilde{v} \in V, \qquad \chi(\tilde{v}) \in \begin{cases} [-1, 0), & v \in V_m^- \\ \{0\}, & v \notin V_m \\ (0, 1], & v \in V_m^+ \end{cases} \tag{4.30}$$

and it assigns a signed real weight to the sublanguages $L(\tilde{v}, v)$.

Similar to the measure of regular languages [16] [12] [15], the characteristic vector is denoted as: $\mathbf{X} = [\chi_1 \ \chi_2 \ \cdots \ \chi_n]^T$, where $\chi_j \equiv \chi(v_j)$, is called the $\mathbf{X}$-vector. The j-th element χ_j of $\mathbf{X}$-vector is the weight assigned to the corresponding terminal state v_j. Hence, the $\mathbf{X}$-vector is also called the state weighting vector.

As mentioned before, the marked language $L_m(\Gamma_i)$ consists of both good and bad event strings that, starting from the initial state v_i, lead to V_m^+ and V_m^- respectively. Any event string belonging to the language $L^0(\Gamma_i) = L(\Gamma_i) - L_m(\Gamma_i)$ terminates on one of the non-marked states belonging to $V - V_m$; and L^0 does not contain any one of the good or bad strings. Equivalence classes for regular grammars are now defined in the setting of Myhill-Nerode Theorem [7]. The regular languages $L(\Gamma_i)$ and $L_m(\Gamma_i)$ are expressed as:

$$L(\Gamma_i) = \bigcup_{v_k \in V} L(v_i, v_k) = \bigcup_{k=1}^{n} L(v_i, v_k) \tag{4.31}$$

$$L_m(\Gamma_i) = \bigcup_{v_k \in Q_m} L(v_i, v_k) = L_m^+(\Gamma_i) \cup L_m^-(\Gamma_i) \tag{4.32}$$

where the sublanguage $L(v_i, v_k) \subseteq \Gamma_i$, having the initial state v_i, is uniquely labeled by the terminal state v_k, $k \in \mathcal{I}_\mathcal{Q}$, and $L(v_i, v_j) \cap L(v_i, v_k) = \emptyset \; \forall j \neq k$; and $L_m^+ \equiv \bigcup_{v \in V_m^+} L(v_i, v)$ and $L_m^- \equiv \bigcup_{v \in V_m^-} L(v_i, v)$ are good and bad sublanguages of the marked language $L_m(\Gamma_i)$, respectively. The null sublanguage $L^0(\Gamma_i) = \bigcup_{v \notin V_m} L(v_i, v)$ and $L(\Gamma_i) = L^0(\Gamma_i) \cup L_m^+(\Gamma_i) \cup L_m^-(\Gamma_i)$.

A signed real measure is constructed as $\mu^i : 2^{L(\Gamma_i)} \to \mathbf{R} \equiv (-\infty, +\infty)$ on the σ-algebra $2^{L(\Gamma_i)}$. The construction is exactly equivalent to that for the state-based automata. With the choice of this σ-algebra, every singleton set made of an event string $\omega \in L(\Gamma_i)$ is a measurable set, which qualifies itself to have a numerical quantity based on the above decomposition of $L(\Gamma_i)$ into L^0(null), L^+(positive), and L^-(negative), respectively called null, positive, and negative sublanguages. The event costs are defined below.

Definition 4.4.5 *Given a set P of production rules, the event cost of the regular grammar Γ_i is defined as follows:*
$\tilde{\pi} : \Sigma^* \times V \to [0, 1]$ *such that* $\forall v_i \in V$, $\forall \sigma_j \in \Sigma$, $\forall s \in \Sigma^*$,

$$\tilde{\pi}[\sigma_j, v_i] \equiv \tilde{\pi}_{ij} \in [0, 1); \sum_j \tilde{\pi}_{ij} < 1; \tilde{\pi}[\epsilon, v_i] = 1;$$

$$\tilde{\pi}[\sigma_j, v_i] = 0 \text{ if } \nexists v_i \to \sigma_j v_k \in P; \qquad (4.33)$$

$$\tilde{\pi}[\sigma_j s, v_i] = \tilde{\pi}[\sigma_j, v_i] \; \tilde{\pi}[s, v_k], \text{ where } v_i \to \sigma_j v_k \in P.$$

Definition 4.4.6 *The state transition cost, $\pi : V \times V \to [0, 1)$, of the regular grammar Γ_i is defined as follows:*

$$\forall v_i, v_j \in V, \pi_{ij} = \begin{cases} \sum_{\sigma \in \mathcal{I}_\Sigma} \tilde{\pi}[\sigma, v_i], & \text{if } \{\sigma \in \Sigma : v_i \to \sigma v_j\} \in P \\ 0 & \text{otherwise.} \end{cases} \qquad (4.34)$$

Consequently, the $n \times n$ state transition cost $\mathbf{\Pi}$-matrix is defined as:

$$\mathbf{\Pi} = \begin{bmatrix} \pi_{11} & \pi_{12} & \cdots & \pi_{1n} \\ \pi_{21} & \pi_{22} & \cdots & \pi_{2n} \\ \vdots & \vdots & \ddots & \vdots \\ \pi_{n1} & \pi_{n2} & \cdots & \pi_{nn} \end{bmatrix} \qquad (4.35)$$

Remark 4.4.3 *The variable-based Definition 4.4.5 and state-based Definition 6.2.4 of the event cost function are exactly equivalent and so are definitions of variable-based and state-based transition cost in Definitions 4.4.6 and 6.2.5. Therefore, the same Π-matrix is obtained if the variables are interpreted as states.*

Definition 4.4.7 *The signed real measure μ^i of every singleton string set $S = \{s\} \in 2^{L(\Gamma_i)}$ where $s \in L(v_i, v)$ is defined as $\mu(S) \equiv \tilde{\pi}(s, v_i)\chi(v)$. The signed real measure of the sublanguage $L(v_i, v) \subseteq L(\Gamma_i)$ is defined as*

$$\mu(L(v_i, v)) = \left(\sum_{s \in L(v_i, v)} \tilde{\pi}[s, v_i] \right) \chi(v) \tag{4.36}$$

The signed real measure of the language of a regular grammar Γ_i initialized at a state $v_i \in V$, is defined as:

$$\mu_i \equiv \mu^i(L(\Gamma_i)) = \sum_{v \in \Gamma} \mu^i(L(v, v_i)) \tag{4.37}$$

The language measure vector, $\boldsymbol{\mu} \equiv [\mu_1 \ \mu_2 \ \cdots \ \mu_n]^T$, is called the $\boldsymbol{\mu}$-vector.

Based on the reasoning of the state based approach, it follows that:

$$\mu_i = \sum_j \pi_{ij}\mu_j + \chi_i \tag{4.38}$$

In vector form, $\boldsymbol{\mu} = \boldsymbol{\Pi}\boldsymbol{\mu} + \mathbf{X}$ whose solution is given by:

$$\boldsymbol{\mu} = (\mathbf{I} - \boldsymbol{\Pi})^{-1}\mathbf{X} \tag{4.39}$$

Remark 4.4.4 *The matrix $\boldsymbol{\Pi}$ is a contraction operator and hence $(\mathbf{I} - \boldsymbol{\Pi})$ is invertible. So, the $\boldsymbol{\mu}$-vector in Equation (4.39) is uniquely defined.*

4.5 Language Measure for Linear Grammars

This section extends the concept of language measure to linear grammars (V, P, T, S) that are a generalization of regular grammars [7]. This is accomplished by introducing the notion of *event plane* from the perspective that there exists a specific direction in which an event may occur in a linear grammar.

Definition 4.5.1 *The event mapping $\eta : \Sigma \to \mathbb{Z}$ is a function that maps the event alphabet into the set of integers. Let $\Sigma = \{\sigma_1, \cdots, \sigma_k, \cdots, \sigma_\ell\}$, then*

$$\eta(\sigma_k) = k \tag{4.40}$$

The *event plane* can be viewed as the complex plane itself on which the trajectory of the discrete-event system is reconstructed as the strings are generated. The transitions $S \to \sigma_k v_i$ and $S \to v_i \sigma_k$ transfer the state located at the origin $(0, 0)$, to $(\eta(\sigma_k), 0)$ and $(0, \eta(\sigma_k))$, respectively. Thus, there exists two possible directions in which the same event σ_k may cause transition from the same state v_i. For a linear grammar, events occur along a specific direction that may be either the real axis or the imaginary axis, depending on the representation being right regular or left regular. This concept is clarified by using the notion of the event plane.

An event is denoted by σ if it occurs along the real axis, and by $i\sigma$ if it occurs along the imaginary axis; let the alphabet of real events be named as Σ and the alphabet of imaginary events as $i\Sigma$. Note that Σ and $i\Sigma$ are disjoint finite sets of identical cardinality. Furthermore, each $\sigma_j \in \Sigma$ is uniquely identified with the specific $i\sigma_j \in i\Sigma$.

4.5.1 Construction of Linear Grammar Measure

Several concepts need to be clarified before embarking on the construction of a measure for linear grammars. In finite state machines, a state q_k may be reached from a state q_j through different paths. Similarly, there may exist multiple paths between two states of a linear grammar. Whereas a particular string uniquely determines a path of a given automaton, a single string may be generated by different paths in linear grammars.

Definition 4.5.2 *The path mapping function $\wp : \Sigma \cup i\Sigma \to \Sigma$ generates a string of real events by concatenating all the real events followed by reverse concatenation of the real events corresponding to the remaining imaginary events. For example, a path mapping is: $\wp(\sigma_j \sigma_k i \sigma_\ell \sigma_p i \sigma_q \sigma_r) = \sigma_j \sigma_k \sigma_p \sigma_r \sigma_q \sigma_\ell$.*

It follows from Definition 4.5.2 that, given a path ω in the language, the generated string is obtained by the path mapping $\wp(\omega)$. The objective is to construct a measure of the set of all such paths rather than the measure of strings. This is necessary for a general form of the Myhill-Nerode theorem to hold, which is central to the construction of the linear grammar measure.

Let $\Upsilon \equiv \{x : x = a + ib \text{ and } a \in [0,1], b \in [0,1]\}$, i.e., Υ is the closed unit square on the complex plane C. Let a binary operator $\star : C \times C \to C$ be defined as: $(a + ib) \star (c + id) = ac + ibd \ \forall a, b, c, d \in R$. The identity for the operator $\star$ is $(1 + i)$ since $\forall z \in C, z \star (1 + i) = z$ and if $z = a + ib$ with $a \neq 0$ and $b \neq 0$, then $\exists$ a unique $z^{-1} \in C$ such that $z^{-1} = \frac{1}{a} + i\frac{1}{b}$. That is, $z \star z^{-1} = (1 + i)$. The operator $\star$ can be extended to multi-dimensional cases by $\star : C^{n \times m} \times C^{m \times l} \to C^{n \times l}$ as follows.

If $\mathcal{A} \in C^{n \times m}$, $\mathcal{B} \in C^{m \times l}$, then $\mathcal{A} \star \mathcal{B} = \mathcal{C} \in C^{n \times l}$ in the sense that $c_{ij} = \Sigma_{k=1}^n a_{ik} \star b_{kj}$. Further, if $\mathcal{A} = A_r + iA_{im}$ and $\mathcal{B} = B_r + iB_{im}$ where the pairs (A_r, B_r) and (A_{im}, B_{im}) denote the real and imaginary parts of the matrices $\mathcal{A}$ and $\mathcal{B}$, respectively, it follows that $\mathcal{A} \star \mathcal{B} = A_r B_r + iA_{im}B_{im}$.

The identity for the above $\star$ operation is $(1 + i)I$ where I is the standard $n \times n$ identity matrix. The identity matrix in the $\star$ operation is denoted as

$$\mathcal{I} = (1 + i)I \tag{4.41}$$

where $\mathcal{A} \star \mathcal{I} = \mathcal{I} \star \mathcal{A} = \mathcal{A} \quad \forall \mathcal{A} \in C^{n \times n}$.

Remark 4.5.2 *The inverse of a matrix $\mathcal{A} \in C^{n \times n}$ under the $\star$ operation, if it exists, is given as: $\mathcal{A}^\S = (A_r + iA_{im})^\S = A_r^{-1} + iA_{im}^{-1}$ that is different from the standard inverse $\mathcal{A}^{-1}$. Notice that both real (A_r) and imaginary (A_{im}) parts of the matrix $\mathcal{A}$ must be individually invertible in the usual sense for existence of $\mathcal{A}^\S$.*

In view of the $\star$ operator, definitions of the language measure parameters, χ, $\tilde{\pi}$ and π, are generalized as follows.

Definition 4.5.3 *The characteristic function* $\chi : V \to \Upsilon$ *is defined as*

$$\forall v \in V,\ \chi(v) = k(1+i),\ \text{where} \begin{cases} k \in [-1, 0) & if\ v \in V_m{}^{-} \\ k = 0 & if\ v \notin V_m \\ k \in (0, 1] & if\ v \in V_m{}^{+} \end{cases} \quad (4.42)$$

The characteristic value χ assigns a complex weight to a language $L(v, u)$ that, starting at the variable v, ends at the variable u. A real weight, in the range of -1 to +1, is assigned to each state as it was done in the case of real grammars, and then this weight is made complex by multiplying (in the usual sense) with $1 + i$. Each event can occur either along the real axis or along the imaginary axis on the event plane. The events are defined to be elements of the set $\{1, i\} \times \Sigma$ with $(1, \sigma)$ denoted as σ and (i, σ) denoted as $i\sigma\ \forall \sigma \in \Sigma$. Furthermore, an event $\sigma \in \Sigma$ becomes $i\sigma$ if it occurs along the imaginary axis; and real occurrences are denoted just by σ similar to what was done for real grammars.

Definition 4.5.4 *The event cost of the LCFG* Γ *is defined as a function* $\tilde{\pi} : (\{1, i\} \times \Sigma)^* \times V \to \Upsilon$ *such that* $\forall v_k, v_\ell \in V,\ \forall \sigma_j \in \Sigma,\ \forall \omega \in (\{1, i\} \times \Sigma)^*$,

(1) $\tilde{\pi}[\sigma_j, v_k] \equiv \mathbb{Re}(\tilde{\pi}_{kj}) \in [0, 1);\ \sum_j \mathbb{Re}(\tilde{\pi}_{kj}) < 1;$
(2) $\tilde{\pi}[i\sigma_j, v_k] \equiv \mathbb{Im}(\tilde{\pi}_{kj}) \in [0, 1);\ \sum_j \mathbb{Im}(\tilde{\pi}_{ij}) < 1;$
(3) $\tilde{\pi}[\sigma_j, v_k] = i$ *if* $\nexists v_k \to \sigma_j v_\ell \in P;$
(4) $\tilde{\pi}[i\sigma_j, v_k] = 1$ *if* $\nexists v_k \to v_\ell \sigma_j \in P;$
(5) $\tilde{\pi}[\epsilon, v_\ell] = 1 + i;$
(6) $\tilde{\pi}[\tau\omega, v_k] = \tilde{\pi}[\tau, v_k] \star \tilde{\pi}[\omega, v_\ell]$
     where $\tau \in \{1, i\} \times \Sigma;\ \omega \in (\{1, i\} \times \Sigma)^*;$ *and*
     $v_k \to \tau v_l$ *if* $\tau \equiv \sigma$ *and* $v_k \to v_\ell \tau$ *if* $\tau \equiv i\sigma$

Definition 4.5.5 *The state transition cost of the LCFG* Γ *is defined as a function* $\pi : V \times V \to \Upsilon$ *such that* $\forall v_k, v_j \in V$,

$$\pi_{kj} \equiv \sum_{\substack{\sigma_\ell \in \Sigma: \\ \exists v_k \to \sigma_\ell v_j \in P}} \mathbb{Re}(\tilde{\pi}[\sigma_\ell, v_k]) + i \sum_{\substack{\sigma_\ell \in \Sigma: \\ \exists v_k \to v_j \sigma_\ell \in P}} \mathbb{Im}(\tilde{\pi}[k, l])$$

$$(4.43)$$

and $\pi_{kj} = 0$ *if* $\{\sigma \in \Sigma : v_k \to \sigma v_j$ *or* $\sigma \in \Sigma : v_k \to v_j \sigma\} \cap P = \emptyset$. *Consequently, the* $n \times n$ *complex-valued state transition cost* $\mathbf{\Pi}$*-matrix is defined as:*

$$\mathbf{\Pi} = \begin{bmatrix} \pi_{11} & \pi_{12} & \cdots & \pi_{1n} \\ \pi_{21} & \pi_{22} & \cdots & \pi_{2n} \\ \vdots & \vdots & \ddots & \vdots \\ \pi_{n1} & \pi_{n2} & \cdots & \pi_{nn} \end{bmatrix} \quad (4.44)$$

Definition 4.5.6 *Let Γ_k be a linear grammar, initialized at a state $v_k \in V$. The complex measure μ^k of every singleton string set $\Omega = \{\omega\} \in 2^{L(\Gamma_k)}$ is defined as: $\mu(\Omega) \equiv \tilde{\pi}(\omega, v_k) \star \chi(v)$. Then, the complex measure of the sublanguage $L(v_k, v) \subseteq L(\Gamma_k)$ of all strings terminated at the state $v \in V$ is defined as:*

$$\mu^k(L(v_k, v)) = \left(\sum_{\omega \in L(v_k, v)} \tilde{\pi}[\omega, v_k] \right) \star \chi(v) \qquad (4.45)$$

Complex measure of the language of the linear grammar Γ_k is defined as:

$$\mu_k \equiv \mu^k(L(\Gamma_k)) = \sum_{v \in \Gamma} \mu^k(L(v_k, v)) \qquad (4.46)$$

The language measure vector, denoted as $\mu = [\mu_1 \ \mu_2 \ \cdots \ \mu_n]^T$, is called the μ-vector.

4.5.3 Computation of the μ-vector

This subsection presents a procedure to formulate the complex measure of paths in the language $L(\Gamma_k)$.

$$L_k = \left(\cup_j \sigma_k{}^j L_j \right) \cup_k \varepsilon_k$$

where the null event ε_k is defined as

$$\varepsilon_k = \begin{cases} \epsilon, & \text{if self loop at } v_i \\ \emptyset, & \text{otherwise} \end{cases}$$

The above expression formalizes the fact that the set of paths from a state v_k is exactly equal to the union of the sets of paths obtained by looking at the first event and then considering all possible legal paths thereafter. Hence, if the first event is σ_ℓ and the current state changes to v_j, then the set of all paths thereafter is exactly equal to L_j. The expression is structurally identical to that given for $DFSA$ in [16] [15] with the understanding that the event $\sigma_\ell{}^j$ can be either real or imaginary. Hence,

$$\begin{aligned} \mu_k \equiv \mu^k(L_k) &= \mu^k \left(\left(\cup_j \sigma_k{}^j L_j \right) \cup_k \varepsilon_k \right) \\ &= \mu^k \left(\cup_j \sigma_k{}^j L_j \right) + \mu^k(\varepsilon_k) \\ &= \sum_j \mu^k \left(\sigma_k{}^j L_j \right) + \chi(v_k) \\ &= \sum_j \pi_{kj} \star \mu^j(L_j) + \chi(v_k) \end{aligned}$$

$$(4.47)$$

The first three steps in Eq. (4.47) follow from the fact that if the first symbol for two paths is different, then the paths cannot be identical. However, the strings generated may still be the same. The fourth step follows from property (6) of the $\tilde{\pi}$ function in Definition 4.5.4. The final step trivially follows from the definition of the measure in 4.5.6. In vector form, the complex measure $\boldsymbol{\mu} \equiv [\mu_1 \ \mu_2 \ \cdots \ \mu_n]^T$ is given by

$$\mu(L) = \boldsymbol{\Pi} \star \mu + \mathbf{X} = (\mathcal{I} - \boldsymbol{\Pi})^{\S} \star \mathbf{X}$$
$$= \left((\mathbf{I} - \mathbb{R}e\boldsymbol{\Pi}) + i(\mathbf{I} - \mathbb{I}m\boldsymbol{\Pi}) \right)^{\S} \star \mathbf{X}$$
$$= (\mathbf{I} - \mathbb{R}e\boldsymbol{\Pi})^{-1}\mathbb{R}e\mathbf{X} + i(\mathbf{I} - \mathbb{I}m\boldsymbol{\Pi})^{-1}\mathbb{I}m\mathbf{X}$$

where $\mathbb{R}e$ and $\mathbb{I}m$ refer to the real and imaginary parts of the matrices, respectively; and existence of the matrix inverses is guaranteed by the following conditions:

$$\sum_j \mathbb{R}e(\tilde{\pi}_{kj}) < 1; \quad \sum_j \mathbb{I}m(\tilde{\pi}_{kj}) < 1 \ \forall k.$$

A simple example is presented in the next section to illustrate how the complex μ is computed for an *LCFG*.

4.5.4 Example 1

Let a language L generate all strings of the type $\{a^k b^k : k \geq 0\}$ over the alphabet $\Sigma = \{a, b\}$. The non-regular language L can be generated by the grammar $\{v \rightarrow avb|\epsilon\}$ that can be rewritten as: $v_1 \rightarrow av_2$; and $v_2 \rightarrow v_1 b$. The resulting $\tilde{\boldsymbol{\Pi}}$-matrix is obtained as

$$\tilde{\boldsymbol{\Pi}} = \begin{bmatrix} p & 0 \\ 0 & iq \end{bmatrix}$$

where the parameters p and q can be identified from the experimental time series data of the system dynamics [17]. The $\boldsymbol{\Pi}$-matrix is then obtained as

$$\mathbb{R}e\boldsymbol{\Pi} = \begin{bmatrix} 0 & p \\ 0 & 0 \end{bmatrix} \text{ and } \mathbb{I}m\boldsymbol{\Pi} = \begin{bmatrix} 0 & 0 \\ q & 0 \end{bmatrix}.$$

Assigning characteristic values (i.e., weights) of the two states v_1 and v_2 to be $\chi_1(1+i)$ and $\chi_2(1+i)$ respectively, the complex measure vector is evaluated as:

$$\mu(L) = \left\{ \begin{array}{c} (\chi_1 + p\chi_2) + i\chi_1 \\ \chi_2 + i(\chi_2 + q\chi_1) \end{array} \right\}$$

4.6 Unobservablility in Discrete Event Systems

The generated language $L(G_i)$ for a $DFSA$ $G_i \equiv (Q, \Sigma, \delta, q_i, Q_m)$, defined in Section 4.3, is the set of all accepted strings by the $DFSA$ G_i with $Q_m = Q$ and the initial state $q_i \in Q$. Since $L(G_i)$ is a prefix-closed language for every $i \in \mathcal{I}_Q$, a change in the initial state from q_i to q_j with $i \neq j$, leads to a different generated language $L(G_j)$. Thus, it is possible to define different prefix closed languages $L(G_i)$, one for each initial state q_i. In the union of these languages $\bigcup_{i \in \mathcal{I}_Q} L(G_i)$, it may become impossible to distinguish between identical symbol sequences generated from different initial states. This difficulty is alleviated by labelling each string with the index (or *color*) of its initial state. Note that the empty string ϵ has no color.

Definition 4.6.1 *For a given $DFSA$ having the state set Q and symbol alphabet Σ, the color alphabet is defined to be the state index set $\mathcal{I}_Q$. Then, the unique label j associated with the initial state q_j of a string $t \in \bigcup_{i \in \mathcal{I}_Q} L(G_i)$ is called the color of the string t.*

The properties of the color alphabet $\mathcal{I}_Q$ are summarized below:

- There exists a bijective mapping between $\mathcal{I}_Q$ and Q
- $\mathcal{I}_Q \bigcap \Sigma = \emptyset$
- $j\epsilon = j \;\; \forall j \in \mathcal{I}_Q$
- Each symbol in $\mathcal{I}_Q$ has zero string length, i.e., $|j| = 0, \; \forall \, j \in \mathcal{I}_Q$

Definition 4.6.2 *For a given $DFSA$ G_i with the initial state $q_i \in Q$, the ith color language associated with the color alphabet $\mathcal{I}_Q$ is defined to be*

$$L_i^c = iL(G_i) \bigcup \{\epsilon\} \tag{4.48}$$

and the ith total color language is defined to be

$$\mathcal{L}_i^c = i\Sigma^* \bigcup \{\epsilon\} \tag{4.49}$$

Remark 4.6.1 *The ith color language is obtained from the generated language $L(G_i)$ by prefixing all non-null strings with the color of the starting state of the strings, namely i. Note that string lengths do not change due to the presence of the zero-length color. Furthermore, if the state transition function δ in the the $DFSA$ $G_i \equiv (Q, \Sigma, \delta, q_i, Q_m)$ is a total function, then $L_i^c = \mathcal{L}_i^c$.*

Definition 4.6.3 *The complete color language of a $DFSA$ with the color alphabet $\mathcal{I}_Q$ is defined to be*

$$\boldsymbol{L}^c = \bigcup_{i \in \mathcal{I}_Q} L_i^c \tag{4.50}$$

and the complete total color language with the color alphabet $\mathcal{I}_Q$ is defined to be

$$\mathcal{L}^c = \bigcup_{i \in \mathcal{I}_Q} \mathcal{L}_i^c \tag{4.51}$$

Remark 4.6.2 *The complete color language $\boldsymbol{L}^c$ is the (disjoint) union of color languages and hence the problem of distinguishing identical symbol sequences starting from different states is alleviated. In general, $\boldsymbol{L}^c \subseteq \mathcal{L}^c$ with the equality holding if and only if $L(G_i) = \Sigma^* \; \forall \, i \; \in \mathcal{I}_Q$.*

Remark 4.6.3 *For every non-empty (colored) string $s \in \boldsymbol{L}^c$, there exists $j \in \mathcal{I}_Q$, $t \in \Sigma^*$ such that $s = jt$.*

Definition 4.6.4 *The (possibly noncommutative) binary operation $\circ : \boldsymbol{L}^c \times \boldsymbol{L}^c \longrightarrow \mathcal{L}^c$ is defined as*

$$\forall s \in \boldsymbol{L}^c, \; s \circ \epsilon = \epsilon \circ s = \; s$$
$$and \; \forall \; s, \tilde{s} \in \boldsymbol{L}^c - \emptyset, \; s \circ \tilde{s} = (it) \circ (\tilde{i}\tilde{t}) \equiv it\tilde{t} \; \text{ for some } t, \tilde{t} \in \Sigma^* \tag{4.52}$$

Remark 4.6.4 *In general, $\circ$ is not a closed operation. That is, for some $s, \tilde{s} \in \boldsymbol{L}^c$, $s \circ \tilde{s}$ may not be an element of $\boldsymbol{L}^c$. If $s \circ \tilde{s} \in \boldsymbol{L}^c$, then s and $\tilde{s}$ are said to be compatible. It should be noted that compatibility of s and $\tilde{s}$ implies the color of the terminating state of s is the same as the color of the starting state of $\tilde{s}$. That is, if $s \in L_i^c$ and $\tilde{s} \in L_j^c$, compatibility of s with $\tilde{s}$ implies $s \circ \tilde{s} \in L_i^c$.*

In Definition (4.6.4), the operation $\circ$ can be viewed as a special concatenation. In analogy to the extension of string concatenation to language concatenation, the following extension is introduced.

Definition 4.6.5 *The extended binary operation $\circ : 2^{\boldsymbol{L}^c} \times 2^{\boldsymbol{L}^c} \longrightarrow 2^{\mathcal{L}^c}$ is defined as*

$$\forall L \subseteq \boldsymbol{L}^c, \; L \circ \epsilon = \epsilon \circ L = \; L$$
$$and \; \forall L, \widetilde{L} \subseteq \boldsymbol{L}^c, \; L \circ \widetilde{L} = \bigcup_{s \in L} \bigcup_{\tilde{s} \in \widetilde{L}} \{s \circ \tilde{s}\} \tag{4.53}$$

Lemma 4.6.1 *For any $s \in \boldsymbol{L}^c$ with $|s| = k \in \{1, 2, 3, \cdots\}$, $\exists$ unique colored strings $s_1, s_2, \cdots, s_k$ with $s_i \in \boldsymbol{L}^c$ such that $s_1 \circ s_2 \circ \cdots \circ s_k = s$.*

Proof. Let $s = i_1\sigma_1 \cdots \sigma_k$ with $i_1 \in \mathcal{I}_Q$, $\sigma_j \in \Sigma$. This implies $s \in L^c(G_{i_1})$ and

$$q_{i_1} \xrightarrow{\sigma_1} q_{i_2} \xrightarrow{\sigma_2} \cdots \xrightarrow{\sigma_{k-1}} q_{i_k} \xrightarrow{\sigma_k} q_{i_{k+1}} \tag{4.54}$$

Hence, by defining $s_k \equiv i_k\sigma_k$, one can construct $s_k \in L_{i_k}^c$ and it follows from Definition (4.6.4) that $s_1 \circ s_2 \circ \cdots \circ s_k = s$. Uniqueness results from the deterministic property of the language $L(G_i)$.

Corollary 4.6.1 *For any given nonempty $s \in L^c$, $\exists$ unique $j \in \mathcal{I}_Q$, $\sigma \in \Sigma$ and $\tilde{s} \in L^c$ such that $s = (j\sigma) \circ \tilde{s}$.*

Proof. The proof follows directly from Lemma 4.6.1.

Definition 4.6.6 *The color map $c : \mathcal{L}^c \longrightarrow \mathcal{I}_Q \bigcup \{\epsilon\}$ is defined as*

$$\forall s \in \mathcal{L}^c, \quad c(s) = \begin{cases} \epsilon & \text{if } s = \epsilon \\ i \text{ if } s = i\sigma \text{ for some } i \in \mathcal{I}_Q, \ \sigma \in \Sigma^* \end{cases} \tag{4.55}$$

and $c(s)$ denotes the color of the string s.

Definition 4.6.7 *The observation map $p : L^c \longrightarrow \mathcal{L}^c$ is defined as follows:*

$$p(\epsilon) = \epsilon,$$
$$\forall s \in L^c - \{\epsilon\}, \ p(s) = p(s_1 \circ \cdots \circ s_k) = s_{i_1} \circ \cdots \circ s_{i_j} \circ \cdots \circ s_{i_r} \tag{4.56}$$

where $i_j \in \{1, \cdots, k\} \ \forall j \in \{1, \cdots, r\}$ and $p(s)$ denotes the observed string for any given string $s \in L^c$.

Remark 4.6.5 *The following facts hold in view of Definitions 4.6.6 and 4.6.7:*

- *The observed starting state of a string $s \in L^c$ has the color $c(p(s))$.*
- *Since $s_{i_j}, s_{i_{j+1}}$ in Definition (4.6.7) are, in general, not compatible, it is possible that $p(s) \notin L^c$ for a given string $s \in L^c$.*

Definition 4.6.8 *A string $s \in L^c$ is said to be completely unobservable if $p(s) = \epsilon$.*

Definition 4.6.9 *The observed language $\mathcal{O}_p$ for a DFSA with respect to a given observation map $p(\cdot)$, is defined to be the image $\boldsymbol{Im}(p)$ of the observation map p, i.e.,*

$$\mathcal{O}_p = p(L^c) \tag{4.57}$$

Remark 4.6.6 *It follows from the Definition (4.6.9) that the observed language $\mathcal{O}_p \subsetneqq \mathcal{L}^c$ and is not necessarily a subset of $\boldsymbol{L}^c$.*

The observation map completely specifies unobservability of the plant automaton. The two simplest cases of unobservability are presented below.

Definition 4.6.10 *A DFSA plant is said to have state-dependent regular unobservability if*

$$\forall \, s \equiv s_1 \circ s_2 \circ \cdots s_k \ \in \boldsymbol{L}^c \, ,$$
$$p(s) = p(s_1 \circ s_2 \circ \cdots s_k) = p(s_1) \circ p(s_2) \circ \cdots p(s_k) \tag{4.58}$$

A plant is said to have state-independent regular unobservability if, in addition,

$$p(i\sigma) = p(j\sigma) \ \forall \sigma \in \varSigma, \ \forall i, j \in \mathcal{I}_Q \tag{4.59}$$

Remark 4.6.7 *Regular state-dependent unobservability in Definition 4.6.10 can be specified by marking certain transitions in the graph of the DFSA as unobservable. Note that, in general, the same event may be observable at one state and unobservable at another state. When this possibility is precluded, the situation is state-independent regular unobservability. The state-independent regular unobservability is a special case of state-dependent regular unobservability; both cases have been jointly referred to as regular unobservability in the sequel. For regular unobservability, the observation map is completely specified by defining an event string in terms of unit-length strings $\sigma \in \boldsymbol{L}^c$. However, the map $p(\cdot)$ can be more complicated. For example, consider a single state DFSA with its language $L(G_1) = \varSigma^*$, where $\varSigma = \{\sigma\}$, and the observation map is defined by the condition that the symbols in the odd number positions of any string are not observed. Then,*

$$1\sigma \longmapsto \epsilon$$
$$1\sigma\sigma \longmapsto 1\sigma$$
$$1\sigma\sigma\sigma \longmapsto 1\sigma$$
$$1\sigma\sigma\sigma\sigma \longmapsto 1\sigma\sigma$$
$$1\sigma\sigma\sigma\sigma\sigma \longmapsto 1\sigma\sigma$$
$$1\sigma\sigma\sigma\sigma\sigma\sigma \longmapsto 1\sigma\sigma\sigma$$
$$and \ so \ on$$

Therefore, $p(\cdot)$ cannot be defined by specifying its value on the unit length string σ. Interestingly, in this particular example, $\mathcal{O}_p$ is still a regular language, namely $\varSigma^*$, which may not be the case in general.

Definition 4.6.11 *The phantom language* $\mathfrak{U}_p$ *of a DFSA plant with respect to a given observation map p is defined to be the kernel ker(p), i.e.,*

$$\mathfrak{U}_p = \{s \in \boldsymbol{L}^c \mid p(s) = \epsilon\} \tag{4.60}$$

Remark 4.6.8 *It follows from Definition 4.6.11 that the phantom language is a sublanguage of* $\boldsymbol{L}^c$, *but need not be a regular sublanguage. However, for regular unobservability, it will be shown that the phantom language is indeed a regular sublanguage of* $\boldsymbol{L}^c$.

Definition 4.6.12 *For a given DFSA* G_i *and an observation map p, the phantom automaton* $\mathcal{P}(G_i)$ *is defined to be a subautomaton of* G_i *such that the language* $L_i^\varrho \subseteq L(G_i)$, *generated by* $\mathcal{P}(G_i)$, *consists of completely unobservable strings, i.e.,* $p(s) = \epsilon \ \forall \ s \ \in iL_i^\varrho$ *(see Definition 4.6.8). Therefore,*

$$L_i^\varrho = \bigcup_{j \in \mathcal{I}_Q} L_{i,j}^\varrho \tag{4.61}$$

where $L_{i,j}^\varrho \equiv \{s \in L_i^\varrho \mid s \text{ terminates on state } q_j\}$.

Lemma 4.6.2 *For regular unobservability, the phantom automaton denoted by* $\mathcal{P}(G_i)$ *of a DFSA* G_i *is given by Algorithm 1.*

Algorithm 1: Derivation of Phantom Automaton $\mathcal{P}(G_i)$

---

input   : $DFSA\ G_i, \mathcal{I}_Q, \text{p}$
output : $DFSA\ \mathcal{P}(G_i)$
begin
      for $i \in \mathcal{I}_Q$ do
          for $\sigma_j \in \Sigma$ do
              if $p(i\sigma_j)$ is defined $AND\ p(i\sigma_j) \neq \epsilon$ then
                  Delete transition σ_j from state q_i
              end
          end
      end
      Set the initial state of the automaton to q_i
end

Proof. Algorithm 1 gives an automaton with initial state q_i. Denote it by A_i and the language generated by it by $L(A_i)$. Obviously, $L(A_i) \subseteq \Sigma^*$. If $s \in L(A_i)$, then

$$p(is) = p(i_1\sigma_1) \circ p(i_2\sigma_2) \circ \cdots p(i_k\sigma_k) \text{ [Definition 4.6.10]}$$
$$= \epsilon \circ \cdots \circ \epsilon$$
$$= \epsilon$$

which implies $s \in L_i^\varrho$. Conversely, if $s \in L_i^\varrho$, then

$$p(is) = \epsilon = p(i_1\sigma_1) \circ p(i_2\sigma_2) \circ \cdots p(i_k\sigma_k)$$
$$\Rightarrow p(i_j\sigma_j) = \epsilon \; \forall \; j \; \in \{1, \cdots, k\} \; \Rightarrow s \; \in L(A_i)$$

Remark 4.6.9 *Lemma 4.6.2 is true only for regular unobservability. It follows that L_i^ϱ is a regular sublanguage of $L(G_i)$. Thus, $\mathfrak{U}_p = \bigcup_{i\in\mathcal{I}_Q} iL_i^\varrho$ is a regular sublanguage of L^c.*

Next the measure of the language generated by the ith phantom automaton is defined in the sense of the measure construction presented in Section 4.3. Denoting the measure of L_i^ϱ as μ_i^ϱ, the equivalent matrix form is:

$$\boldsymbol{\mu^\varrho} = (\mathbf{I} - \boldsymbol{\mathcal{P}(\Pi)})^{-1} \mathbf{X} \tag{4.62}$$

where $\boldsymbol{\mu^\varrho} \equiv [\mu_1^\varrho \; \mu_2^\varrho \; \cdots \; \mu_n^\varrho]^T$; and the phantom transition cost matrix $\boldsymbol{\mathcal{P}(\Pi)}$ is defined to be the transition cost matrix of the DFSA $\mathcal{P}(G_i)$ (see Definition 6.2.5 in Section 4.3).

Remark 4.6.10 *The inverse $[\mathbf{I} - \mathcal{P}(\boldsymbol{\Pi})]^{-1}$ exists since $\mathcal{P}(\boldsymbol{\Pi})$ is elementwise non-negative and bounded by $\boldsymbol{\Pi}$; and $\boldsymbol{\Pi}$ is a contraction operator implying that $\mathcal{P}(\boldsymbol{\Pi})$ is a contraction as well.*

Definition 4.6.13 *The ith color component $z_i : 2^{L^c - \{\epsilon\}} \rightarrow 2^{L(G_i)}$ is defined as:* $\forall \; L \equiv \{s_1, s_2, \cdots, s_k, \cdots\} \subseteqq L^c - \{\epsilon\}$,

$$z_i(L) = z_i(\{s_1, s_2, \cdots, s_k, \cdots\})$$
$$= \{z_i(\{s_1\}), z_i(\{s_2\}), \cdots, z_i(\{s_k\}), \cdots\} \tag{4.63}$$

where

$$z_i(\{jt\}) = \begin{cases} \{\epsilon\}, & \text{if } i \neq j \\ \{t\}, & \text{if } i = j \end{cases} \quad \text{and } z_i(\emptyset) = \emptyset$$

Now a signed real measure $\vartheta : 2^{L^c} \rightarrow \mathbf{R} \equiv (-\infty, +\infty)$ is constructed on the σ-algebra 2^{L^c}, similar to the construction of measure in Section 4.3. With the choice of this σ-algebra, every singleton set made of a string $s \in L^c$ is a measurable set, which qualifies itself to have a numerical quantity based on the following construction.

$$\forall L \subseteqq \boldsymbol{L^c}, \quad \vartheta(L) = \sum_{i \in \mathcal{I}_Q} \mu^i\big(z_i(L - \{\epsilon\})\big) \tag{4.64}$$

The set function ϑ is a well-defined signed measure that satisfies the following conditions:

- $\vartheta(\emptyset) = 0$
- Finiteness and countable additivity of ϑ inherited from those of μ
- $\vartheta(iL) = \mu^i(L) \ \forall \ L \subsetneqq L(G_i)$.
- $\vartheta(\{\epsilon\}) = \sum_{i \in \mathcal{I}_Q} \mu^i(z_i(\emptyset)) = \sum_{i \in \mathcal{I}_Q} \mu^i(\emptyset) = 0$

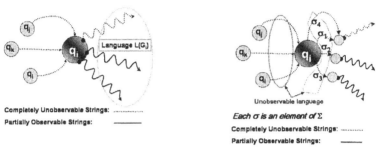

(a) Interpretation of L_i^1 (b) Interpretation of L_i^{11}

Fig. 4.1. Illustration of Unobservable, Partially Observable, and Observable Strings

Construction of the language measure in Section 4.3 facilitates computation of a quantitative measure of the language $L(G_i)$ for all i. This measure is adequate under complete observability and has been successfully used for optimal control under full observation [6]. However, in the generalized case of partial observability, where the observation map is non-trivial, one must consider the fact that strings observed to initiate from a given state may actually start from a different one. A control policy that does not take this effect into consideration may cause colossal errors. For example, the supervisor may enable a controllable event leading to a state of negative weight based on its observation which erroneously indicated that the particular transition will terminate at a state of positive weight. Two sublanguages of $\boldsymbol{L}^c$, denoted by L_i^1 and L_i^{11}, are introduced next to address these issues.

Definition 4.6.14 *The language L_i^1 is the set of all strings which pass through the state q_i with the restriction that the prefix of each string leading to the state q_i is completely unobservable. Formally,*

$$L_i^1 = \left\{ s \in \boldsymbol{L}^c \mid s = s_1 \circ s_2 \ where \ p(s_1) = \epsilon \ and \ s_2 \in L_i^c \right\} \quad (4.65)$$

Definition 4.6.14 implies that L_i^1 is the set of all strings which pass through the state q_i with the restriction that the prefix of each string leading to the state q_i is completely unobservable. Figure 1 clarifies the notion further. The following lemma makes it exact.

Lemma 4.6.3

$$L_i^1 = \bigcup_{k \in \mathcal{I}_Q} kL_{k,i}^\varrho \circ L_i^c \qquad (4.66)$$

Proof. By Definition 4.6.14 $s \in L_i^1 \Rightarrow s = s_1 \circ s_2$ and $s_2 \in L_i^c \Rightarrow$ s_1 terminates on state q_i which follows from the fact that s_1, s_2 has to be compatible since $s_1 \circ s_2 \in \boldsymbol{L}^c$. Hence, s_1 initiates from a state $q_k \in Q$, terminates on state q_i, and is completely unobservable by definition. Hence from Definition 4.6.11, it follows that $s_1 \in kL_{k,i}^\varrho$. Hence, $s \in L_i^1 \Rightarrow s \in \bigcup_{k \in \mathcal{I}_Q} kL_{k,i}^\varrho \circ L_i^c$. The converse follows with a similar reasoning.

Remark 4.6.11 *It is important to note that L_i^1 is, in general, a superset of the set of strings which are observed to initiate at state q_i. For example, if $s \in L_i^1 \Rightarrow s = s_1 \circ s_2$ and $s_2 = i\sigma \circ rs_r$ where $\sigma \in \Sigma, r \in \mathcal{I}_Q$, $s_r \in L_r^c$ and $p(i\sigma) = \epsilon$, then the string is observed to have initiated from state q_r, as seen in the left hand part of Figure 4.1. The string that takes the dotted path after state q_i is not observed to initiate from q_i (i.e., both q_i and its prefix are missed). A solution to the problem is to consider the particular subset of L_i^1 having only those strings for which at least the first transition after q_i is observable.*

Definition 4.6.15 *The language L_i^{11} is defined to be the subset of L_i^1 such that, for any string in L_i^{11}, the first transition after q_i is unobservable. Formally,*

$$\forall\, s \in L_i^{11},\ s = s_1 \circ s_2 \Rightarrow s = s_1 \circ i\sigma \circ s_r \qquad (4.67)$$

where $s_1 \in L_{k,i}^\varrho$, $\sigma \in \Sigma, s_r \in L^c(G_r)$ *for some* $r, k \in \mathcal{I}_Q$, *and* $p(i\sigma) = \epsilon$

Lemma 4.6.4

$$L_i^{11} = \left(\bigcup_{k \in \mathcal{I}_Q} kL_{k,i}^\varrho \right) \circ \left(\bigcup_{\sigma_r^j \in \Sigma_i^\varrho} i\sigma_r^j L(G_j) \right) \qquad (4.68)$$

where Σ_i^ϱ is the set of unobservable events in Σ, which are defined from state q_i; and $\sigma_r^j \in \Sigma_i^\varrho$ such that there exists a transition σ_r^j in the phantom automaton $\mathcal{P}(G_i)$, with $\delta(q_i, \sigma_r^j) = q_j$.

Proof. Following the right hand part of Figure 4.1 and Definition 4.6.15 of L_i^{11}, it follows that
$p(i\sigma_r) = \epsilon$ if and only if $\sigma_r \in \Sigma_i^\varrho \subseteqq \Sigma$. Hence, $s \in L_i^{11} \Rightarrow s \in$ $\left(\bigcup_{k \in \mathcal{I}_Q} kL_{k,i}^\varrho \right) \circ \left(\bigcup_{\sigma_r^j \in \Sigma_i^\varrho} i\sigma_r^j L(G_j) \right)$ and vice versa.

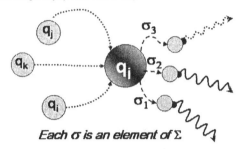

Completely Unobservable Strings: ....
Completely Observable Strings: – – – – –
Partially Observable Strings: ——————

Fig. 4.2. *Illustration of Unobservable and Partially Observable Strings – Interpretation of* $L_i^1 - L_i^{11}$

Lemma 4.6.5 *The set difference* $L_i^1 - L_i^{11}$ *is the collection of all strings in* L^c *observed to have initiated from state* q_i.

Proof. In view of Remark 4.6.11, $L_i^1 - L_i^{11}$ is the set of all strings in L_i^1 for which the first transition after state q_i is observable; that is, these strings are observed to have initiated from state q_i. Figure 4.2 clarifies the idea. The proof follows from Definitions 4.6.14 and 4.6.15.

4.6.1 Computation of Language measure under Partial Observation

This section presents the computation of the measures of L_i^1, L_i^{11} and $\mathbb{L}_i$ from the closed form expressions that are presented as the following theorems.

Theorem 4.6.1 *The measure* ϑ *of the language* $\boldsymbol{L^1}$ *is expressed in closed form as*

$$\vartheta(L^1) = \mathbf{D}\left([\mathbf{I} - \mathcal{P}(\mathbf{\Pi})]^{-T}\left\{\begin{matrix}1\\ \vdots \\ 1\end{matrix}\right\}\right)[\mathbf{I} - \mathbf{\Pi}]^{-1}\mathbf{X} \qquad (4.69)$$

Proof.

$$\vartheta(L_i^1) = \vartheta\left(\bigcup_{k\in\mathcal{I}_Q} kL_{k,i}^\varrho \circ L^c(G_i)\right)$$

$$= \sum_{k\in\mathcal{I}_Q} \vartheta\left(kL_{k,i}^\varrho \circ L^c(G_i)\right)$$

$$= \sum_{k \in \mathcal{I}_Q} \mu^k \left(L_{k,i}^\varrho L(G_i) \right)$$

$$= \sum_{k \in \mathcal{I}_Q} \left\{ \sum_{\omega \in L_{k,i}^\varrho} \widetilde{\pi}^\varrho [\omega, q_k] \right\} \mu^i (L(G_i))$$

$$= \sum_{k \in \mathcal{I}_Q} \left\{ \sum_{\omega \in L_{k,i}^\varrho} \widetilde{\pi}^\varrho [\omega, q_k] \right\} \mu_i$$

$$= \left\{ \mathbf{D} \left([\mathbf{I} - \mathcal{P}(\mathbf{\Pi})]^{-\mathbf{T}} \left\{ \begin{matrix} 1 \\ \vdots \\ 1 \end{matrix} \right\} \right) \mu \right\}_i$$

$$= \left\{ \mathbf{D} \left([\mathbf{I} - \mathcal{P}(\mathbf{\Pi})]^{-\mathbf{T}} \left\{ \begin{matrix} 1 \\ \vdots \\ 1 \end{matrix} \right\} \right) [\mathbf{I} - \mathbf{\Pi}]^{-1} \mathbf{X} \right\}_i$$

The first equality follows from Definition 4.6.14 and Equation (4.66) in the proof of Lemma 4.6.3. The second equality follows from the fact that $jL_{j,i}^\varrho \circ L^c(G_i)$ and $kL_{k,i}^\varrho \circ L^c(G_i)$ are disjoint sets for $j \neq k$ with $\boldsymbol{L^c}$ being a disjoint union of the color languages. The third equality follows from the discussion after the definition of measure ϑ in Equation (4.64) and also the observation that, given a fixed k, all strings in $kL_{k,i}^\varrho \circ L^c(G_i)$ begin from state q_k. The fourth inequality follows from the proof of Theorem 4.3.1 in Section 4.3. The term $\widetilde{\pi}^\varrho$ is analogous to the event cost function (see Definition 6.2.4) for the phantom automaton (see Definition 4.6.11). The fifth equality simply abbreviates $\mu^i(L(G_i))$ by μ_i. In the sixth equality, the expression within the curly braces is a matrix expression that follows directly from Equation (4.26) in Section 4.3. The final equality replaces the expression for μ from Equation (6.10) in Section 4.3 and the result in matrix form is in Equation (4.69).

Remark 4.6.12 *In case of complete observability,* $\boldsymbol{\mathcal{P}(\Pi)} = \mathbf{0}$ *and then,* $\vartheta(L_i^{11}) = \mu^i(L(G_i)) \equiv \mu_i.$

Theorem 4.6.2 *The closed form expression for* $\vartheta(L_i^{11})$ *is given as*

$$\vartheta(L^{11}) = \mathbf{D} \left([\mathbf{I} - \mathcal{P}(\mathbf{\Pi})]^{-T} \left\{ \begin{matrix} 1 \\ \vdots \\ 1 \end{matrix} \right\} \right) \mathcal{P}(\mathbf{\Pi}) [\mathbf{I} - \mathbf{\Pi}]^{-1} \mathbf{X} \qquad (4.70)$$

Proof.

$$\vartheta(L_i^{11}) = \vartheta\left(\left(\bigcup_{k\in\mathcal{I}_Q} kL_{k,i}^{\varrho}\right) \circ \left(\bigcup_{\substack{\sigma_r^j\in\Sigma_i^{\varrho}\\ \delta(q_i,\sigma_r^j)=q_j}} i\sigma_r^j L(G_j)\right)\right)$$

$$= \sum_{k\in\mathcal{I}_Q} \vartheta\left(\left(kL_{k,i}^{\varrho}\right) \circ \left(\bigcup_{\substack{\sigma_r^j\in\Sigma_i^{\varrho}\\ \delta(q_i,\sigma_r^j)=q_j}} i\sigma_r^j L(G_j)\right)\right)$$

$$= \sum_{k\in\mathcal{I}_Q} \mu^k\left(\left(L_{k,i}^{\varrho}\right)\left(\bigcup_{\substack{\sigma_r^j\in\Sigma_i^{\varrho}\\ \delta(q_i,\sigma_r^j)=q_j}} \sigma_r^j L(G_j)\right)\right)$$

$$= \sum_{k\in\mathcal{I}_Q} \left\{\sum_{\omega\in L_{k,i}^{\varrho}} \widetilde{\pi}^{\varrho}[\omega, q_k]\right\}\mu^i\left(\bigcup_{\substack{\sigma_r^j\in\Sigma_i^{\varrho}\\ \delta(q_i,\sigma_r^j)=q_j}} \sigma_r^j L(G_j)\right)$$

$$= \sum_{k\in\mathcal{I}_Q} \left\{\sum_{\omega\in L_{k,i}^{\varrho}} \widetilde{\pi}^{\varrho}[\omega, q_k]\right\} \sum_{j\in\mathcal{I}_Q}\left\{\sum_{\substack{\sigma\in\Sigma_i^{\varrho}\\ \delta(q_i,\sigma)=q_j}} \widetilde{\pi}^{\varrho}[\sigma, q_i]\right\}\mu_j$$

$$= \sum_{k\in\mathcal{I}_Q} \left\{\sum_{\omega\in L_{k,i}^{\varrho}} \widetilde{\pi}^{\varrho}[\omega, q_k]\right\} \sum_{j\in\mathcal{I}_Q} \mathcal{P}(\mathbf{\Pi})_{ij}\mu_j$$

$$= \sum_{k\in\mathcal{I}_Q} \left\{\sum_{\omega\in L_{k,i}^{\varrho}} \widetilde{\pi}^{\varrho}[\omega, q_k]\right\} (\mathcal{P}(\mathbf{\Pi})\mu)_i$$

$$= \left\{\mathbf{D}\left([\mathbf{I} - \mathcal{P}(\mathbf{\Pi})]^{-\mathbf{T}}\left\{\begin{matrix}1\\ \vdots\\ 1\end{matrix}\right\}\right)\mathcal{P}(\mathbf{\Pi})\mu\right\}_i$$

$$= \left\{\mathbf{D}\left([\mathbf{I} - \mathcal{P}(\mathbf{\Pi})]^{-\mathbf{T}}\left\{\begin{matrix}1\\ \vdots\\ 1\end{matrix}\right\}\right)\mathcal{P}(\mathbf{\Pi})[\mathbf{I} - \mathbf{\Pi}]^{-1}\mathbf{X}\right\}_i$$

The first equality in the proof follows from Definition 4.6.15. The next three equalities follow exactly in the same way as in Theorem 4.6.1. In the fifth equality, μ_j denotes $\mu^i(L(G_i))$ and it follows from the fact that, for $r_1 \neq r_2$, $\sigma_{r_1}^j$ and $\sigma_{r_1}^j$ are disjoint languages. In the sixth equality,

$$\left\{ \sum_{\substack{\sigma \in \Sigma_i^e \\ \delta(q_i,\sigma)=q_j}} \widetilde{\pi}^e[\sigma, q_i] \right\}$$ has been replaced with $\mathcal{P}(\Pi)_{ij}$ which follows directly

from the definition of the transition cost function (Definition 6.2.5) in Section 4.3. The seventh equality simply replaces $\sum_{j \in \mathcal{I}_Q} \mathcal{P}(\Pi)_{ij}\mu_j$ by $(\mathcal{P}(\Pi)\mu)_i$ and the eighth equality follows directly from Equation (4.26) in Section 4.3. The final expression follows from Equation (6.10) in Section 4.3.

Theorem 4.6.3 *The closed form expression for the measure of the language under partial observation is obtained as follows:*

$$\vartheta(L^1 - L^{11}) = \mathbf{D}\left([\mathbf{I} - \mathcal{P}(\Pi)]^{-T} \begin{Bmatrix} 1 \\ \vdots \\ 1 \end{Bmatrix} \right) [\mathbf{I} - \mathcal{P}(\Pi)][\mathbf{I} - \Pi]^{-1}\mathbf{X} \quad (4.71)$$

Proof. The proof follows from Theorems 4.6.1 and 4.6.2.

Remark 4.6.13 *Under complete observability, the set of strings observed to initiate from any given state q_i is the set of strings that do actually start from q_i, i.e., under complete observability, $L^1 = L(G_i)$ and $L^{11} = \emptyset$. Theorem 4.6.3 is consistent with this observation because the following identities hold under complete observability: $\mathcal{P}(\Pi) = 0$ and hence $\vartheta(L^1 - L^{11}) = \mu^i(L(G_i))$.*

4.6.2 Example 2

In this example, the concept of the langauge measure is illustrated under loss of observability of ceratin events at the supervisory level. Figure 4.3 shows a deterministic finite state automaton (DFSA) with three states and an alphabet $\Sigma = \{\alpha, \sigma\}$, where the event α is unobservable. The state transition cost matrix is as follows.

$$\Pi = \begin{bmatrix} 0 & 0.8 & 0.1 \\ 0.1 & 0 & 0.8 \\ 0.8 & 0.1 & 0 \end{bmatrix} \text{ and } \mathcal{P}(\Pi) = \begin{bmatrix} 0 & 0 & 0.1 \\ 0.1 & 0 & 0 \\ 0 & 0.1 & 0 \end{bmatrix};$$

and assuming $\mathbf{X} = [1 \ 0 \ -1]^T$, the language measures are given by

$$\mu = \begin{bmatrix} 0.7287 \\ -0.2834 \\ -0.4453 \end{bmatrix}; \ \vartheta(L^1) = \begin{bmatrix} 0.8097 \\ -0.3149 \\ -0.4948 \end{bmatrix}; \text{ and } \vartheta(L^1 - L^{11}) = \begin{bmatrix} 0.8592 \\ -0.3959 \\ -0.4633 \end{bmatrix}.$$

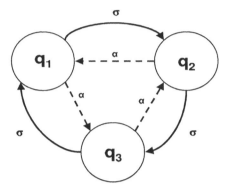

Fig. 4.3. *DFSA with unobservable transitions shown by dotted lines*

4.7 Quantification of Unobservability

Given a plant model, described by the regular language $L(G_i)$, this section quantifies the effects of event unobservability at the supervisory level in terms of the following exact short sequence of monoids [13].

$$0 \longrightarrow \mathfrak{U}_p \overset{\imath}{\hookrightarrow} L^c \overset{p}{\longrightarrow} \mathcal{O}_p \longrightarrow 0 \qquad (4.72)$$

where $\imath$ is the inclusion map and p is the natural projection onto $\mathcal{O}_p$. Exactness of the sequence results from the facts that $\imath$ is injective, p is surjective and **ker** p = **Im** $\imath$. If this exact sequence is split, unobservability has no effect because each observed string would have a unique mapping back to an element of L^c. In general, the map p can be used to define an equivalence relation on L^c as follows:

$$\forall \ s_i, \ s_j \in L^c, ((p(s_i) = p(s_j)) \ \Rightarrow \ (s_i \sim s_j)) \qquad (4.73)$$

This induces a partition $\mathcal{P}$ on L^c with at most countable number of equivalence classes, and each equivalence class of $\mathcal{P}$ has the property that every string in the equivalence class maps to the same element of $\mathcal{O}_p$. The cardinality bound on the number of equivalence classes is due to the fact that the number of strings in L^c itself is at most countable. Moreover, it follows that $\forall \ s_j \in \mathcal{O}_p, \ p^{-1}(s_j) = \mathcal{P}_k$ for some k.

Theorem 4.7.1 *If the phantom language $\mathfrak{U}_p$ (see Definition 4.6.11) is a context free language (CFL), then $\forall \ s_j \in \mathcal{O}_p, \ p^{-1}(s_j)$ is a regular sublanguage of L^c.*

Proof. Let $s_j = j\sigma_1\sigma_2\cdots\sigma_N$ for some $j \in \mathcal{I}_Q$. Then, $p^{-1}(s_j) = (\mathfrak{U}_p\sigma_1 \circ \mathfrak{U}_p\sigma_2 \circ \ldots \circ \mathfrak{U}_p\sigma_N \circ \mathfrak{U}_p) \bigcap L^c$. Closure of context free languages under concatenation implies $(\mathfrak{U}_p\sigma_1 \circ \mathfrak{U}_p\sigma_2 \circ \ldots \circ \mathfrak{U}_p\sigma_N \circ \mathfrak{U}_p)$ is a context free language.

Then, closure under intersection implies $p^{-1}(s_j)$ is context free as well. Detailed proofs of the closure properties are given in [7].

Corollary 4.7.1 *In case of regular unobservability, $\forall\ s_j\ \in \mathcal{O}_p$, the inverse image $p^{-1}(s_j)$ is a regular sublanguage of L^c.*

Definition 4.7.1 *The unobservability index β_{cum} for a given discrete event plant model under a specified unobservability situation is defined as follows:*

$$\beta_{cum}\ \in \mathbf{R}\quad such\ that$$
$$\beta_{cum}\ =\ \sum_{s\in\ \mathcal{O}_p}\left|\ \sup_{\tau\in p^{-1}(s)}\vartheta(\tau)\ -\ \inf_{\tau\in p^{-1}(s)}\vartheta(\tau)\right|$$

The unobservability index β_{cum} quantifies the maximum cumulative error one may incur from the viewpoint of language measure due to the presence of unobservable transitions.

The salient properties of β_{cum} are delineated below.

- $\beta_{cum} \in [0,\infty)$. A specific upper bound is determined from the bounded total variation property of the language measure [15].
- If the observation map p is injective, then $\forall\ s \in \mathcal{O}_p$, $p^{-1}(s)$ is a singleton set and hence $\beta_{cum}\ =\ 0$. This is expected because it follows, from the short exact sequence, that an injective p implies **Im** $\imath$ is trivial, i.e., the only unobservable string is the empty string ϵ.
- $\beta\ =\ 0 \Rightarrow \forall\ s \in\ \mathcal{O}_p$, $\sup_{\tau\in p^{-1}(s)}\vartheta(\tau)\ =\ \inf_{\tau\in p^{-1}(s)}\vartheta(\tau)$. From the bounded total variation of the language measure, it follows that, for $\beta_{cum} = 0$, if $\mathbf{Card}(p^{-1}(s))$ is countably infinite, then $\vartheta(\tau)\ =\ 0 \forall\ \tau \in p^{-1}(s)$, i.e., $p^{-1}(s)$ is a set of strings of zero measure. However, it is possible that $\vartheta(\tau)$ has a nonzero value $\forall\ \tau\ \in p^{-1}(s)$ if $p^{-1}(s)$ is a finite set.

4.7.1 Example 2 Revisited

In the example in section 4.6.2, the observed language $\mathcal{O}_p$ has the regular expression $\mathcal{I}_Q\sigma^*$, where the superscript $*$ indicates Kleene closure operation. Now, the unobservability index β_{cum} is calculated summing the unobservabiliuty β_i of individual subsets.

Considering the set of strings in $p^{-1}(1\epsilon)$, the maximum possible measure of a string is $(0.2)^3 = 0.008$ achieved by the string $\alpha\alpha\alpha$, and the minimum possible measure is -0.2 achieved by the singleton string α and it weight $\chi = -1$. In the calculations that follow, β_i denotes the unobservability calculated from a subset $\mathcal{O}_i \subsetneq \mathcal{O}_p$ such that $\forall s \in \mathcal{O}_i$, $c(s) = 1$ (see Definition 4.6.6). Continuation of this process yields

$$\begin{aligned}
\beta_1 =\ & (0.2^3 + 0.2) + \\
& (0.2 + 0.2^2) \times (0.8 + 0.8^4 + 0.8^7 + \cdots) + \\
& (1 + 0.2^2) \times (0.8^2 + 0.8^5 + 0.8^8 + \cdots) + \\
& (1 + 0.2) \times (0.8^3 + 0.8^6 + 0.8^9 + \cdots) \\
=\ & (0.2^3 + 0.2) + \\
& (0.2 + 0.2^2) \times (0.8/(1 - 0.8^3)) + \\
& (1 + 0.2^2) \times (0.8^2) \times (1/(1 - 0.8^3)) + \\
& (1 + 0.2) \times (0.8^3) \times (1/(1 - 0.8^3)) \\
=\ & 3.2244
\end{aligned}$$

Similarly,

$$\begin{aligned}
\beta_2 =\ & (0.2^2 + 0.2) + \\
& (0.2 + 0.2^2) \times (0.8^3) \times (1/(1 - 0.8^3)) + \\
& (1 + 0.2^2) \times (0.8) \times (1/(1 - 0.8^3)) + \\
& (1 + 0.2) \times (0.8^2) \times (1/(1 - 0.8^3)) \\
=\ & 3.7705
\end{aligned}$$

and

$$\begin{aligned}
\beta_3 =\ & (0.2^2 + 0.2^3) + \\
& (0.2 + 0.2^2) \times (0.8^2) \times (1/(1 - 0.8^3)) + \\
& (1 + 0.2^2) \times (0.8^3) \times (1/(1 - 0.8^3)) + \\
& (1 + 0.2) \times (0.8) \times (1/(1 - 0.8^3)) \\
=\ & 3.4211
\end{aligned}$$

Sum of the three components together yields the unobservability measure

$$\begin{aligned}
\beta_{cum} =\ & 3.2244\ +\ 3.7705\ +\ 3.4211 \\
=\ & 10.416
\end{aligned}$$

Calculation of the unobservability index for a general situation could be numerically complicated mainly due to computational complexity, which is a topic of future research.

4.8 Summary and Conclusions

This chapter has addressed two major research areas in language-based supervisory control.

- Measure of non-regular languages for modelling complex dynamical systems that cannot be efficiently represented by regular languages (equivalently, finite state automata).

- Measure of regular languages for supervisory control synthesis under partial observation (i.e., in the presence of unobservable events at the supervisory level).

Regarding the first research area, this chapter presents a quantitative measure of non-regular languages [2], generated by *linear context free grammars (LCFG)* which belong to the low end of Chomsky hierarchy [7] [9]. It shows that the measure of regular languages proposed in [16] [15] [12] can be obtained by its generating regular grammar, without referring to states of the automaton. Then, the chapter extends the signed real measure to a complex measure for the class of non-regular languages, generated by *LCFG*. The extended language measure is potentially applicable to quantitative analysis and synthesis of DES control systems where the plant model of a complex dynamical system is not restricted to be a finite state machine. Future research in this area includes generalization of the measure to a wider class of context free grammars and beyond, with specific interest in Petri nets. In addition, the issue of measuring paths rather than strings needs to be explored in more details.

Regarding the second research area, a new concept has been presented for future research in extension of the quantitative control policy [6] under loss of observability of certain events. This chapter extends the concept, formulation and validation of the signed real measure for regular languages and their sublanguages under loss of observability of certain events. Specifically, it brings out the difference in measures under full observability and partial observability. A closed form solution is obtained for the language measure under partial observation, which converges to the measure under full observation as the set of unobservable events becomes empty. The optimal control policy under full observation [6] need to be updated to accommodate loss of observability as well as functional relationships of language measure with the amount of available information on a given DFSA.

References

1. C.G. Cassandras and S. Lafortune, *Introducrion to discrete event systems*, Kluwer Academic, 1999.
2. I. Chattopadhyay, A. Ray, and X. Wang, *A complex measure of non-regular languages for discrete-event supervisory control*, Proceedings of American Control Conference, Boston, MA, June-July 2004, pp. 5120–5125.
3. V. Drobot, *Formal languages and automata theory*, Computer Science Press, 1989.
4. J. Fu, C.M. Lagoa, and A. Ray, *Robust optimal control of regular languages with event cost uncertainties*, Proceedings of IEEE Conference on Decision and Control, December 2003, pp. 3209–3214.
5. J. Fu, A. Ray, and C.M. Lagoa, *Optimal control of regular languages with event disabling cost*, Proceedings of American Control Conference, Denver, Colorado, June 2003, pp. 1691–1695.

6. J. Fu, A. Ray, and C.M. Lagoa, *Unconstrained optimal control of regular languages*, Automatica **40** (2004), no. 4, 639–648.
7. J. E. Hopcroft, R. Motwani, and J. D. Ullman, *Introduction to automata theory, languages, and computation, 2nd ed.*, Addison-Wesley, 2001.
8. R. Kumar and V. Garg, *Modeling and control of logical discrete event systems*, Kluwer Academic, 1995.
9. J. C. Martin, *Introduction to languages and the theory of computation, 2nd ed.*, McGraw-Hill, 1997.
10. J.O. Moody and P.J. Antsaklis, *Supervisory control of discrete event systems using petri nets*, Kluwer Academic, 1998.
11. P.J. Ramadge and W.M. Wonham, *Supervisory control of a class of discrete event processes*, SIAM J. Control and Optimization **25** (1987), no. 1, 206–230.
12. A. Ray and S. Phoha, *Signed real measure of regular languages for discrete-event automata*, Int. J. Control **76** (2003), no. 18, 1800–1808.
13. J.J. Rotman, *Advanced modern algebra, 1st ed.*, Prentice Hall, 2002.
14. W. Rudin, *Real and complex analysis, 3rd ed.*, McGraw-Hill, New York, 1987.
15. A. Surana and A. Ray, *Signed real measure of regular languages*, Demonstratio Mathematica **37** (2004), no. 2, 485–503.
16. X. Wang and A. Ray, *A language measure for performance evaluation of discrete-event supervisory control systems*, Applied Mathematical Modelling **28** (2004), no. 9, 817–833.
17. X. Wang, A. Ray, and A. Khatkhate, *On-line identification of language measure parameters for discrete event supervisory control*, Proceedings of 42nd IEEE Conference on Decision and Control (Maui, Hawaii), December 2003, pp. 6307–6312.

Part II

Engineering and Software Applications of
Language Measure and Supervisory Control

5

Discrete Event Supervisory Control of a Mobile Robotic System

Xi Wang[1], Peter Lee[2], Asok Ray[3], and Shashi Phoha[4]

[1] The Pennsylvania State University xxw117@psu.edu
[2] The Pennsylvania State University cfl106@psu.edu
[3] The Pennsylvania State University axr2@psu.edu
[4] The Pennsylvania State University sxp26@psu.edu

Summary. As an application of the theory of Discrete Event Supervisory (DES) control presented in Chapters 1 and 2, this chapter addresses the design of a robotic system interacting with a dynamically changing environment. The work, reported in this chapter, encompasses the disciplines of control theory, signal analysis, computer vision, and artificial intelligence. Several traditional important methods are first reviewed to substantiate the DES control approach in the design of a mobile robotic system. Design and modelling of the behavior-based mobile robotic system are presented in details. The plant automaton model G of the robotic system is identified by making use of the available sensors and actuators. Then, a DES controller is synthesized based on the data collected from experimental scenarios. Through these experiments, performance of the robotic DES control system is quantitatively evaluated in terms of the language measure μ for both the unsupervised and supervised robotic systems. It is shown that the language measure μ can indeed be used as a performance index in the design of optimal DES control policies for higher level mission planning for behavior-based mobile robotic systems.

Key words: Discrete Event Supervisory Control, Robot Control, Computer Vision, Language Measure

5.1 Supervisory Control System design

This section addresses the design of optimal supervisory control for a behavior-based mobile robotic system. Often a world model is used by the planner to make the most appropriate sequence of actions for the agent. This approach suffers from poor scalability and requires frequent replanning of robot schedules due to uncertainty in sensing and action, and changes in the environment. Traditional top-down *planner-based* or *deliberative* approaches have been shown to be inadequate for responding, in real time, to abrupt changes in the physical environment [14] [6] [7].

Purely reactive behavior-based bottom-up approaches embed the agent's control strategy into a collection of preprogrammed parallel condition-action pairs with minimal internal state and no search; hence, real-time performance can be achieved [14]. The modularly designed component behaviors are activated in parallel, producing various commands to corresponding actuators. These behaviors represent low-level continuous control and have been successful in making local incremental decisions. However, within this paradigm, it is often difficult to execute the higher level tasks by solely using artificial intelligence (AI) symbolic planning techniques because no global state of the world is available. To address this issue within the subsumption architecture [5], behaviors are represented as finite state machines that are augmented with a set of input and output channels to provide a communication link to higher and lower layers. The lower level behaviors (e.g., *walking, attraction to a goal, repulsion from an obstacle*) are active for most of the time, and they trigger or deactivate the high level behaviors (e.g., *wandering around, avoidance*). The higher level behaviors are designed to "subsume" or override the output of the lower level. Once the layered network is completed, it remains fixed since the arbitration scheme is hardwired in the architecture.

The discrete event approach has been proposed to the robotics research community by Košecká [12] for navigation of mobile robots with various tasks defined as discrete states, e.g., *moving, steer_away, path_following*. Similarly, Feddema et al [8] use the discrete states: *Search, Rotate, Backup*, and *Track* for the line following problem. Discrete event systems are appropriate for modelling of complex systems consisting of many interacting components operating in a dynamically changing environment driven by abrupt changes (also known as discrete events). In this sense, discrete event approach is similar to the AI symbolic planning approaches in which the overall robotic system is hierarchically structured into three levels: *symbolic planning, reactive behaviors, and low level servoing*. On the other hand, the discrete event approach provides a theoretically sound background and a formal treatment in the design of autonomous mobile robots. Considering the inherently discrete nature of a robot's behavior, asynchronous interruptions from the environment, and discrete extracted features from sensors, it is reasonable to model a mobile robot in the discrete event framework. Therefore, the coordinated control system of an autonomous mobile robot is formulated as a supervisory decision and control problem for discrete event systems.

5.1.1 System architecture

The design features of the behavior-based robotic system are delineated as follows:

1. *Fast response.* A robot needs local intelligence to perform autonomously. Consequently, the discrete event supervisor (DES) that runs locally is hosted in the robot platform.
2. *Flexibility.* For different mission objectives, it is only necessary to redesign a supervisor without any alteration in the robot's configuration.
3. *Resource Saving.* For status monitoring, only high level symbolic representation (e.g., *searching, grasping*) of the robot functions is communicated to the remote site.
4. *Robustness and Fault Tolerance.* Robustness to detrimental effects of component faults is achieved by a discrete-event switching mechanism. Upon detecting a failure, an appropriate supervisor is switched on at its proper state.
5. *Coordination.* The robot has the capability to coordinate its behavior with a set of robots. In other words, the robotic system is scalable.

Figure 5.1 shows the architecture of the behavior-based multi-layer robotic system that integrates the finite state automaton model of event driven discrete event dynamics with the finite-dimensional model of time driven continuous dynamics through appropriately defined continuous/discrete (C/D) and discrete/coninuous (D/C) interfaces. The discrete event supervisor mod-

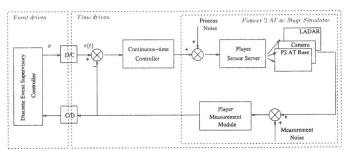

Fig. 5.1. DES behavior based robotic system architecture

ule (DESM) communicates with the robot's continuous time-varying control module(CTCM) by the sending and receiving of a set of discrete events. DESM is designed to be independent of the underlying physical process dynamics and provides a mechanism to interact with CTCM; the controller unit in DESM is allowed to plug and play. CTCM consists of three blocks: a continuously-varying time-driven controller; C/D interface; and D/C interface. CTCM connects to Player via a standard TCP socket for sending and receiving formatted messages that encode up-to-date sensor readings and continuously varying commands of reference signals, respectively, at 10 Hz. The control strategy is event-driven, in which CTCM (in particular, the C/D block) generates discrete events based on the continuous sensor data received from the Player. The discrete events are transmitted to DESM in symbolic form. DESM reacts immediately by sending out a a discrete event command back to CTCM according to the currently loaded supervisor. After receiving a particular event from the DES controller, CTCM sends out a set of continuous reference signals to Player for completion of this behavior to maneuver the robot accordingly. The robot continues executing this behavior until the sensor data triggers the occurrence of a new event. In this architecture, DES control strategy for both the robot and its simulator are designed and exercised. The hardware integration of a self-customized robot is given in the next section.

5.1.2 Hardware integration

The Pioneer 2 AT robot[5], as shown in Figure 5.2, is equipped with a SICK LMS200 laser range finder for obstacle avoidance and distance measurement. The laser range finder provides depth information for a 180° field of view with an angular resolution of 0.5° and an accuracy of 1 cm($\pm$15 percent). The robot uses a SONY EVI-D30 pan-tilt camera in conjunction with a Sensoray 311

---

[5] Pioneer 2 AT is a product of ActivMedia Inc.

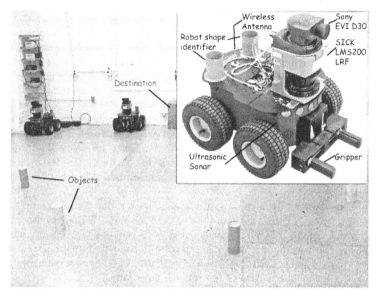

Fig. 5.2. Robotic system lab and Pioneer 2 AT specifications

PC104 frame grabber for object recognition and tracking. For communication between the robot and a remote computer for the purpose of monitoring, a Lucent Technologies WaveLan 11 Mbps radio ethernet (2.4GHz) is employed. In practice, a bandwidth of up to 2Mbps is achieved. An Advantech© on-board computer powered by a Transmeta Crusoe Processor TM5400 500MHz CPU performs all real-time computations. It has 310MB memory including 256MB PC133 RAM and 64MB flash memory, and a 20GB hard disk. These devices are powered by three 12 Volt sealed lead acid batteries; DC/DC converters are used to provide appropriate power to various devices. The major actuators available on the robots are motors for wheel drive, gripper, and camera.

5.1.3 Stage simulator and robot kinematics

Stage[6] simulates a population of mobile robots, sensors, and environmental objects in a two-dimensional bitmapped environment [9]. Stage enables 1) rapid design and development of controllers that eventually drive the robots in real time; 2) robot experiments without access to the hardware and environments. Stage emulates, with sufficient fidelity, a set of devices including sonar, laser range finders, visual color segmenters, and a versatile mobile robot

---

[6] Player/Stage is developed at University of Southern California under GNU Public License. It is available at http://playerstage.sourceforge.net/

base with an odometer. The robot configurations in experiment and simulation are shown in Figures 5.3 and 5.5, respectively. Robot sensor readings and simulated sensor readings are displayed in Figures 5.4 and 5.6, respectively, with largely similar environment settings. In these figures, the yellowish brown color beams represent sonar readings and light blue colored areas bounded by dark blue dotted lines represent the laser readings. Real-time sonar readings are much more noisy than simulated sonar readings. The boundary of the laser readings is shown by straight lines in the areas without obstacles where simulated robot moves. (Note that the robot experiment area is much smaller than that of the simulated robot.) All devices in Stage are accessible through Player's standard interfaces, as if they were real hardware. In design of the robotic system, identical control software is used on both the robot and Stage simulator.

The behaviors of the robot and the simulated robot are comparable, as explained below. The full configuration of the Pioneer 2 AT robot is described as a directed point, i.e., an ordered pair $\mathbf{q} = (x, y, \theta) \in \mathcal{C} = \mathbb{R}^2 \times S^1$, where (x, y) are the coordinates with respect to a global reference frame F_g; the current orientation θ is with respect to x-axis; and $\mathcal{C}$ is the robot configuration space. Nonholonomic motion of the robot is constrained by:

$$-\dot{x}\sin\theta + \dot{y}\cos\theta = 0 \qquad (5.1)$$

which limits the robot to two degree of freedom (DOF) and hence a motion command of this robot can be fully characterized by two motion parameters only at any given configuration. The kinematics of Pioneer 2 AT can then be governed by the following equation:

$$\begin{pmatrix} \dot{x} \\ \dot{y} \\ \dot{\theta} \end{pmatrix} = \begin{pmatrix} \cos\theta & 0 \\ \sin\theta & 0 \\ 0 & 1 \end{pmatrix} \begin{pmatrix} v \\ w \end{pmatrix} \qquad (5.2)$$

where the steering input w controls the angular velocity $\dot{\theta}$ and the driving input v controls the linear velocity along the direction of the wheel. To achieve higher fidelity, continuously varying system identification of the Pioneer 2 AT robot motion dynamics is performed and the resulting model, along with sonar, laser range finder, vision camera, and gripper, is used in the simulation experiments. The experiments were conducted with pseudo-random inputs in both time and speed under the following constraints: (i) $2 \text{ sec} \leq t_d \leq 10 \text{ sec}$; (ii) $-300 \text{ mm/s} \leq v \leq 300 \text{ mm/s}$; (3) $-80 \text{ deg/s} \leq \omega \leq 80 \text{ deg/s}$, where t_d is the time duration in which the robot runs at constant speed; v and ω are the linear and angular velocities of the robot, respectively. By analyzing the experimental data, a multivariable ARX model was found to be inaccurate to capture the nonlinear dynamics of this nonholonomic robot, therefore the subspace technique for system identification of MIMO systems was applied using the MATLAB toolbox provided in [15]. The resulting state space model, having the state vector $x = [v \ w \ \dot{v} \ \dot{w}]$, is given as:

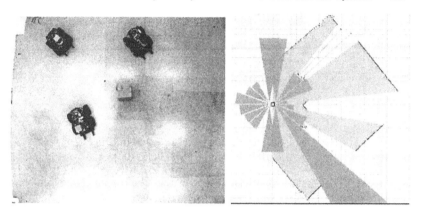

Fig. 5.3. Robot configuration in experiment

Fig. 5.4. Sensor readings in experiment

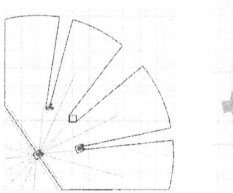

Fig. 5.5. Robots configuration in simulator

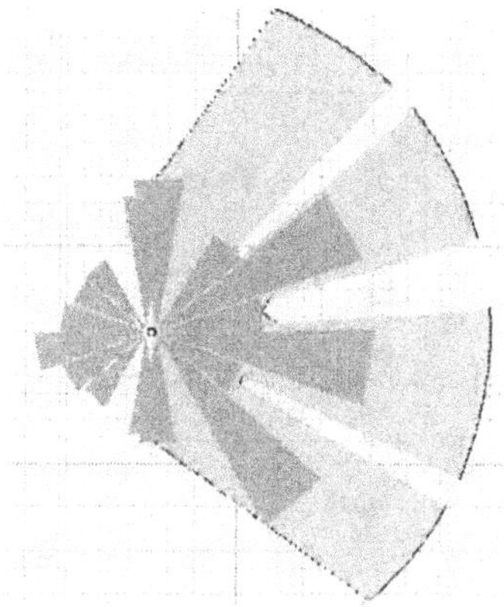

Fig. 5.6. Sensor readings in simulator

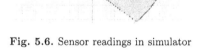

$$x_{n+1} = Ax_n + Bu_n \qquad (5.3)$$
$$y_n = Cx_n \qquad (5.4)$$

$$A = \begin{bmatrix} 0.8579 & 0.0958 & -0.3488 & -0.0118 \\ -0.0012 & 0.6882 & -0.1532 & 0.0540 \\ 0.1415 & 0.3104 & 0.6644 & -0.5791 \\ -0.0893 & 0.2350 & 0.3603 & 0.6374 \end{bmatrix}$$

$$B = \begin{bmatrix} -0.3081 & -0.0037 \\ -0.0032 & 0.6524 \\ 0.4281 & -0.2781 \\ -0.1809 & -0.4719 \end{bmatrix}$$

$$C = \begin{bmatrix} -0.4808 & 0.2368 & -0.5129 & -0.2716 \\ 0.0138 & 0.4664 & 0.0547 & 0.2910 \end{bmatrix}$$

Figure 5.7 for system identification shows that the results of the random walk experiment on the robot system are in good agreement with those predicted by the model. The variations of the model predictions from the actual measured data are largely due to environmental disturbances (e.g., floor friction and unflatness).

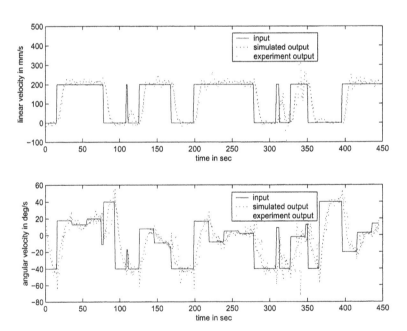

Fig. 5.7. System identification of the Pioneer 2 AT robot

5.1.4 Software design and integration

The robot's on-board computer runs on Linux Redhat 7.3 kernel 2.4.7-10. The Player robotic development platform is employed for both the robot system and the simulator Stage. Player is a robot device server which manages the robot's sensors and actuators. The robot continuous time control block in Figure 5.1 talks to Player through local port following the publishing/subscribing mechanism. Both the DES supervisor and the robot's continuous time controller run locally on the on-board computer. If coordination is needed, only the DES supervisor exchanges discrete events possibly attached with certain continuous and/or discrete parameters with coordinator (or other robots). Table 5.1 provides an approximate distribution of CPU usage for a set of applications that are directly related to the robot operation.

Table 5.1. CPU usage of various applications

Applications	Version	CPU usage
ACTS	2.0	30%
Player	1.3.1	$\leq 1\%$
CTCM	1.3.1	0.4%
DESM	1.3.1	$\leq 0.2\%$
sshd	3.4	20%

ActiveMedia Color Tracking Software (ACTS) is the program for color segmentation and connected components analysis at run time. It consumes about 30% of CPU time for object detection. Although sshd (abbreviation for *secure shell demon*) is used for remote login to the robot for more detailed monitoring, it may not be required in actual experiments.

5.1.5 Design of robot behaviors

A behavior is a set of processes involving sensing and action against the environment. Behaviors can be designed at a variety of levels of abstraction. In general, they are made to be higher than the robot's atomic actions, e.g., *turn left by 30 degree*, and they extend in time and space. Commonly implemented behaviors include: *go home, search object, avoid obstacle, pick up object*, etc. In the discrete event setting, a behavior is a controllable event. The union of these behaviors is the controllable event set Σ_c. The set of uncontrollable events Σ_u is the set of all possible responses of the robot during its interactions with its environment. It consists of the results of the robot's behaviors, including a successful or unsuccessful completion of the controllable events (e.g., *reach a target, grab an object successfully*) and an interruption of the robot's current behavior (e.g., *find an object, detect obstacle*).

The C/D and D/C blocks in Figure 5.1 are the interfaces between the discrete event dynamics and continuous time dynamics of the robot system.

The D/C block is a map $\phi : \Sigma \to \mathbb{R}^m$ that converts each controllable event into the continuous reference signal as follows:

$$r(t) = \phi(\sigma) \qquad t_k \le t < t_{k+1} \tag{5.5}$$

where $\sigma \in \Sigma$ is the most recent controllable event and t_k is the time of the k-th discrete event occurrence. The C/D block is a map ψ that converts the state space $\mathbf{x}$ of the continuous time system into the alphabet of discrete events Σ. An event σ_τ is generated at time $t = \tau$ if there exist $\epsilon, \delta > 0$, such that $\forall 0 < \epsilon < \delta$

$$h(\mathbf{x}, \mathbf{u}; \tau) = 0, \text{ and } h(\mathbf{x}, \mathbf{u}; \tau - \epsilon) \ne 0 \tag{5.6}$$
$$\sigma_\tau = \psi(\mathbf{x}) \tag{5.7}$$

Equation (5.6) defines a closed set Ω and an event is generated whenever the state trajectory enters Ω from outside for the first time. That is, if there is a segment of trajectory $\mathbf{x}$ during $[t_1, t_2]$ that satisfies Equation 5.6, then the event is defined to be occurred at t_1. The C/D block (also known as the event generator) is completely specified by the pair (h, ψ) where the function h is defined in Equation (5.6).

Based on all possible (limited by its hardware) robot behaviors Σ_c and all possible response Σ_u of the robot during its interaction with the environment, the alphabet of discrete events Σ is identified as listed in Table 5.2. Note that $\Sigma_u \equiv \Sigma_u \cup \Sigma_c$, where Σ_c and Σ_u are given by

$$\Sigma_c = \{a, d, g, h, i, s, v\}$$
$$\Sigma_u = \{A, c, C, l, o, p, P, q, Q, T, w, W, x, X, y\}$$

It should be noted that no two events can be triggered exactly at the same time instant.

5.1.6 Visual servoing

The rigid body motion of the mobile frame F_m is attached to the robot and changes over time with respect to a fixed spatial reference frame F_r, as shown in Figure 5.8 (a). Let $V_{F_m}(t) = [x, y, z]^T \in \mathbb{R}^3$ be the position vector of the origin of frame F_m from the origin of reference frame F_r and the orientation angle θ is defined in the counter-clockwise sense about the z-axis, as shown in Figure 5.8 (a). Now, suppose a camera mounted on the Pioneer 2 AT robot which is facing downward with a tilt angle $\phi > 0$ and it is above the ground plane by distance h_c, as shown in Figure 5.8 (b). The x-axis of camera coordinate frame F_c is chosen to be the optical axis of the camera, the y-axis of F_c coincides with y-axis of F_m, and the optical center of the camera coincides with the origins of both F_m and F_c. Then the kinematics of a point $p_c = [x, y, z]^T$ attached to the camera frame F_c is given in the instantaneous camera frame by [13]:

Table 5.2. The discrete event set Σ for Pioneer 2 AT robot

Symbol	Description	Controllable
a	approach the object	$\checkmark$
A	avoid obstacle successfully	
c	reach goal with an object	
C	find an object but gripper full	
d	drop an object	$\checkmark$
g	grab an object	$\checkmark$
h	return to home	$\checkmark$
i	ignore the current observed target	$\checkmark$
l	lost the target	
o	obstacle ahead	
p	drop an object successfully	
P	fail to drop an object	
q	grab an object successfully	
Q	fail to grab an object	
T	find goal with an target	
s	search recognizable target	$\checkmark$
v	avoid obstacle	$\checkmark$
w	reach target without an object	
W	find object 1	
x	lost the goal	
X	find target 2	
y	lost an object	

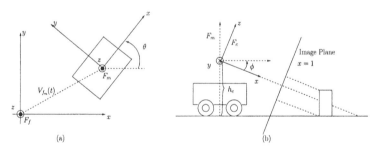

(a)   (b)

Fig. 5.8. (a) Model of Pioneer 2 AT robot; (b) The side-view of P2AT with a camera facing downward at a tilt angle $\phi > 0$

$$\begin{pmatrix} \dot{x} \\ \dot{y} \\ \dot{z} \end{pmatrix} = \begin{pmatrix} \cos\phi \\ 0 \\ \sin\phi \end{pmatrix} v + \begin{pmatrix} -y\cos\phi \\ z\sin\phi + x\cos\phi \\ -y\sin\phi \end{pmatrix} w \qquad (5.8)$$

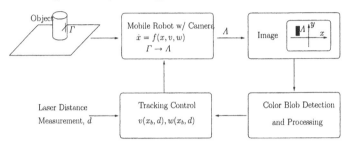

Fig. 5.9. The closed-loop vision-guided navigation system for the Pioneer 2 AT robot

It is well-known that the system in Equation 5.2 does not satisfy Brockett's condition [4][7]. For this reason, visual servoing based switched controllers are developed since these have the potential to provide better overall performance. The major advantage of visual feedback control is that the accuracy of the closed-loop performance can be made relatively insensitive to calibration errors and nonlinearities of the open-loop system. It is the only source of information that the robot can use in a feedback loop to navigate to an identified target in an unknown environment as it does not process any knowledge of the current positions of itself and the target. Moreover, localization techniques, if used, are in general very noisy, erroneous, and computation intensive, making it very difficult to have robust *approach* and *grab* behaviors. Figure 5.9 shows a block diagram of the visual servoing system. In the image plane, a feature parameter space, denoted by $f \in \mathcal{F}_I$, is adopted to represent those parameters that describe objects being tracked. In the robotic system, a colored destination and a set of same-size different-colored cylindrical objects are being tracked by the real-time color blob detection software ACTS. The output of ACTS is further regrouped by Player into a vector of blobs representing the feature parameters of any detected objects. Each blob is parameterized by $B(y_b, z_b, A_b)$, where (y_b, z_b) are the coordinates of the blob B centroid in the image plane and A_b is the area of the blob B.

The idea of visual servoing is to regulate a particular error function to zero. In the robotic system, visual servoing attempts to keep a feature staying at the desired image plane coordinate. To do that, the wheel motion controller and camera motion controller are engaged accordingly so that the blob B stays approximately stationary in both y and z directions. Therefore, the x value along the optical axis of the camera in Equation (5.8), i.e., the depth of the blob, is not of particular interest. Since dynamics effects are not prevalent for steering tasks at low speed and nominal driving conditions, kinematic models

---

[7] A nonholonomic systems with more degrees of freedom than controls cannot be stabilized by continuously differentiable, time invariant, state feedback control laws

have been used to derive the control laws. Visual servoing starts to take over robot motion control whenever an object is detected by ACTS. It operates in two distinct stages. The first stage is to steer the robot towards the object by reducing the angular difference between the robot and the object, i. e., $\Delta\theta \approx 0$. The second stage then moves the robot to the desired point while keeping centering the blob in the image plane. The design of robot motion control is described below.

Robot motion control – first stage: On the image plane, the centroid, (y_b, z_b), of a blob represents the angular offset between the robot and the object. A proportional control law is applied to steer the robot accordingly:

$$w = -k_w \text{sign}(y_b)|y_b| \qquad (5.9)$$

The linear velocity v is determined by the blob's z-coordinate, denoted as z_{b0}, on the image plane at the instant of the robot's detecting an object. During the search, the camera tilt angle ϕ is set to be zero so that if z_{b0} is negative, the object is close to the robot; and if z_{b0} is positive, the object is far away from the robot. Based on this experience, the linear velocity v is given by:

$$v = \begin{cases} v_0 + k_{v1}\text{sign}(z_{b0})|z_{b0}|, & z_{b0} > 0 \text{ and } |y_b| > y_{th} \\ v_0 + k_{v2}\text{sign}(z_{b0})|z_{b0}|, & z_{b0} \le 0 \text{ and } |y_b| > y_{th} \\ v_1, & |y_b| \le y_{th} \end{cases} \qquad (5.10)$$

To keep track of the object that the robot is approaching, the camera is set to tilt according to z_b so that the feature (blob) is maintained in the camera's field of view once the *approach* behavior is triggered. The camera motion control command is sent through a RS232 connection between the on-board computer and the camera by the VISCA protocol. A proportional control law is applied to the camera motor.

$$w_t = -k_c \text{sign}(z_b)|z_b| \qquad \phi \le \phi_{cr} \qquad (5.11)$$

where ϕ_{cr} is the pre-specified camera tilt angle above which visual tracking of object becomes unstable due to smaller blob and worse illumination of the object.

Robot motion control – second stage: Although visual tracking is no longer available at this stage, the robot is very close to the object and the orientation of the robot with respect to the object is correct. Accordingly, the robot motion control law is given by:

$$v = \begin{cases} v_2, & t \le t_d \\ -v_3, & t > t_d \end{cases} \qquad w = \begin{cases} 0, & t \le t_d \\ -w_1, & t > t_d \end{cases} \qquad (5.12)$$

where t_d is the time duration within which the robot should be able to reach the object. The performance of visual servoing has been experimentally validated in the scenario of a robot's approaches an object, as shown in Figure 5.10. The robot stops if two laser beams between the gripper are obstructed; and the event *reach object* is generated. However, for fault-tolerant design, a time limit t_d is set for the final stage of a robot's approaching the object. If the time limit is expired and the two laser beams are still not obstructed, the robot backs up, rotates away, resets the camera, and generates the *object lost* event. This may happen because of various reasons including device imperfections and malfunctions, or dynamically changing environments.

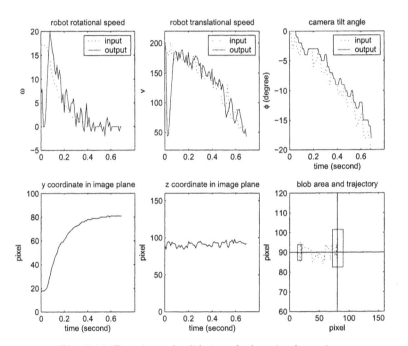

Fig. 5.10. Experimental validation of robot visual servoing

In practice, the blob detection is sensitive to noisy environment, particularly, the illumination condition of the object that the robot is tracking. Therefore, if the blob under tracking is lost, the *object lost* event is generated and reported to the DES controller. In addition, the *approaching* behavior may also be interrupted by the *found obstacle* event generated by another sensor module, for example, the laser ranger finder that and measures the distance.

5.1.7 Obstacle avoidance

Obstacle avoidance is an essential behavior that all mobile robots need to acquire for autonomous navigation in an unknown and (possibly) changing environment. Generally speaking, there are two types of distance-sensor-based obstacle avoidance algorithms: global and local. Equipped with a global map of the operational environment, global obstacle avoidance algorithms can be naturally integrated into the obstacle-free path planning of the mobile robot as applications of Potential Field(PF) method [10] [1] [11] that is briefly described below.

In potential field (PF) methods, a robot is modeled as a particle with electric charge. Each obstacle is modeled as a number of particles on its boundary with the same electric charge as the robot. The goal is modeled as a particle with the opposite electric charge to the robot. An artificial potential function is first defined at the robot's position (x, y).

$$U(x,y) = U_g(x,y) + U_r(x,y) \tag{5.13}$$

where $U_g(x, y)$ is the attractive potential produced by the goal particle at (x, y), and $U_r(x, y)$ is the repulsive potential produced by all the obstacle particles. Under this model, the robot's collision-free path planning at any given location (x, y) is simple: examine the robot neighborhood, and move to the location with the lowest potential field value. In practice, the vector sum of the repulsive force $\mathbf{F}_r(x, y)$ and attractive force $\mathbf{F}_g(x, y)$ on the robot at (x, y) forms a force $\mathbf{F}$ whose direction is then the most appropriate direction that the robot should move on. The primary advantage of potential field methods is that they are extremely fast to compute. However, they typically consider only a small subset of obstacles near the robot [11].

5.1.8 LRF readings segmentation and noise rejection

Sensor readings from laser range finder (LRF) could be noisy, as shown in Figure 5.11. The half circles show the boundary of robot's safe zone beyond which it may collide with an obstacle. A simple filtering method is applied, where the main idea is to subdivide the laser scanner readings in each scan into small segments. (Note: A segment is a set of measurement data points that are close enough to each other and thus are assumed to belong to the boundary of the same object.) Those data points that do not belong to any of the segments are replaced with neighboring data. The set of laser range data measured at $t = t_k$, denoted as $\mathcal{R}_k$, is given as:

$$\mathcal{R}_k = \{\ell_i \mid \ell_i = (r_i, \theta_i), i \in [0, N]\} \tag{5.14}$$

where ℓ_i is the i-th data points with distance r_i and angle θ_i, and N is the total number of data points. The distance between two consecutive points is given by:

$$d(\ell_i, \ell_{i+1}) = \sqrt{r_{i+1}^2 + r_i^2 - 2r_{i+1}r_i \cos(\theta_{i+1} - \theta_i)} \tag{5.15}$$

Let $\ell_i \in \mathcal{C}_j$, where $\mathcal{C}_j$ is the j-th segment of the reading $\mathcal{R}_k$. $\ell_{i+1} \in \mathcal{C}_j$ if

$$d(\ell_i, \ell_{i+1}) \leq c_0 + c_1 \min\{r_i, r_{i+1}\} \tag{5.16}$$

where $c_1 = \sqrt{2(1 - \cos(\theta_{i+1} - \theta_i))}$. The constant c_0 allows an adjustment of the algorithm to the noise level, chosen to be 0.5m in the experiments. The constant c_1 is the coefficient associated with the distance between consecutive points in each scan under the assumption that the two data points have no abrupt change in the distance. A conservative choice is $\min\{r_i, r_{i+1}\}$ that provides flexibility for selection of the cost function. A minimum size of five data points is used in each segment, i.e., $|\mathcal{C}_j| \geq 5, j = 1, 2, \ldots$. Figure 5.12 shows the samples in Figure 5.11 after filtering.

A polar histogram of the filtered laser range data is then developed for the purpose of obstacle avoidance. In implementation, the measurement data points are partitioned into $n = \frac{360°}{\alpha}$, where $\alpha = 5°$ is the resolution of an angular sector. The following computations are made in each sector β_k:

$$\forall k = 1, \ldots, n, \quad \ell_k^{\min} = \min_j \{\ell_j \mid \ell_j \in \beta_k\}, \quad w_k = \begin{cases} \ell_k^{\min} & \ell_k^{\min} \leq d_{\min} \\ 0 & \ell_k^{\min} \leq d_{\min} \end{cases} \tag{5.17}$$

Figure 5.13 shows a typical measurement of laser range data. If there are some sectors that have non-zero w_k, one choice for the next direction of motion of the mobile robot is the middle angle of the largest open area on the histogram plot. Figure 5.14 shows an example of such a histogram, and the robot can move toward about the 300° direction. From experiments, such a polar-histogram-based obstacle avoidance is always active and works in real time.

5.1.9 Vector field histogram

Local obstacle avoidance algorithms generally do not require any global map because decisions are made on sensor readings. There exists a number of algorithms that integrate both global and local information, such as Occupancy Grids and Vector Field Histogram(VFH) family. Vision-based obstacle avoidance algorithms still suffer from poor performance due to real-time constraints and sensitivity to environment lighting conditions. In the robotic system design, a procedure that first performs filtering on noisy sensor(LRF) readings and then applies the VFH+ algorithm for obstacle avoidance is chosen. The VFH+ method [17], which is an improved version of the Vector Field Histogram (VFH) method [3] [2] for real-time, local obstacle avoidance of mobile robots, is implemented in the robotic system. The basic idea of the VFH+ method is that the 2-dimensional map grid is reduced to a 1-dimensional polar histogram H that comprises n angular sectors of width α degrees. The polar histograms constructed over time are always around the robot's momentary location. Each cell in H represents the obstacle density in that particular direction. Based on the obstacle density profile, which includes the masked polar histogram and a cost function, a decision is then made as the safest direction for the robot to move. The VFH+ method consists of four stages

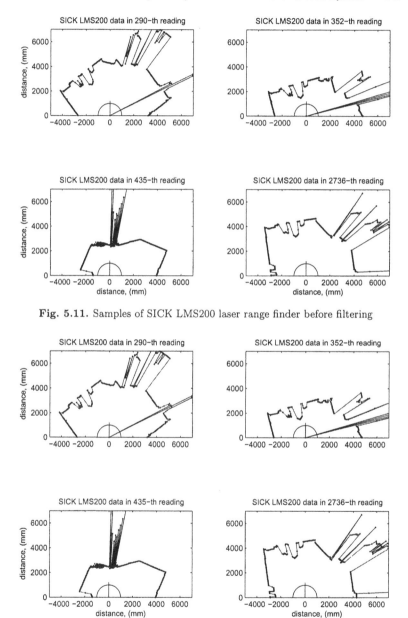

Fig. 5.11. Samples of SICK LMS200 laser range finder before filtering

Fig. 5.12. Samples of SICK LMS200 laser range finder after filtering

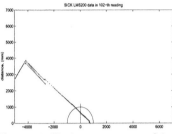

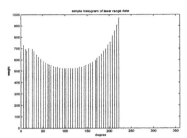

Fig. 5.13. Single measurement of data   **Fig. 5.14.** Histogram of data

for data reduction and computation of the new direction of motion. Details of this method can be found in [17]. During the *searching* behavior, the robot runs at a speed of 0.2m/s as a trade-off between the robot motion and visual blob detection.

5.2 Experimental scenario and parameter identification

As stated early in this chapter, the robot can perform a set of behaviors, including *search, home, avoid, ignore, approach, grab,* and *drop*. The experiment scenario is designed to have the robot collecting colored objects to a pre-specified destination. There are two colored (*green* and *pink*) objects randomly located in the field every time after the robot collects one of them back to the destination. Without any discrete event supervisory control, the robot starts with random *search* and, upon detecting an object, it has two options: *approach* or *ignore* the object. If ignoring the object, it should continue the random search. Otherwise, it should approach the object. Upon successfully reaching the object, it should grab the object and begin the random search to find the destination provided that the object grabbing is successful. If the destination is found, the robot can either *approach* or *ignore*. Once the robot decides to approach the destination and succeeds in reaching the destination, it should drop the object. The robot repeats the same procedures until stopped by a human operator.

Given the above experiment scenario, an automaton model G of the unsupervised robot is formulated as shown in Figure 5.15. A non-blocking [8] supervisor S is synthesized according to the following specification language K.

1. Whenever an object is found, the robot must approach it;
2. The robot approaches the destination if it has grabbed an object and found the destination.

---

[8] A supervisor S is nonblocking if $pr(L(S/G)) \cap L_m(G) = L(S/G)$ (See [16]).

The automaton model S/G of the supervised robot is shown in Figure 5.16. One can easily check that S is controllable with respect to the plant G and uncontrollable event set Σ_{uc}.

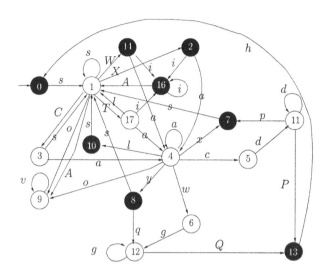

Fig. 5.15. Unsupervised robot automaton model G

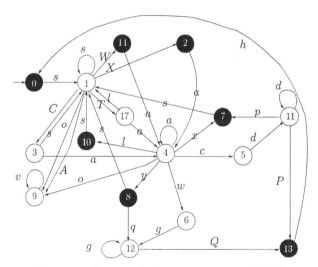

Fig. 5.16. Supervised robot automaton model S/G

Figure 5.17 and Figure 5.18 show the convergence of selected non-zero elements in $\tilde{\Pi}$-matrix obtained in both robot experiments and robot simulation according to the parameter identification procedure described in Chapter 1. One can see the probability of finding object 2 at the supervisor state 1 in the robot experiment is higher than then that in the simulation. This is because realization of random positioning of the objects in simulation is much easier than in real-time experiments. In fact, due to lighting condition, the robot has a higher probability of losing an object, as shown in the figures. So the objects are randomly placed in a much smaller area of the available test field in order for the robot to find them.

Table 5.3. The state set Q of the plant model G and its $\mathbf{X}$-vector

State	Description	χ value
0	robot ready for mission	0.0
1	searching for target	0.0
2	found object 2	0.1
3	fount an object but gripper is full	0.0
4	approaching target	0.0
5	ready to drop object at destination	0.0
6	ready to grab object	0.0
7	drop an object successfully	0.8
8	grab an object successfully	0.4
9	avoiding obstacle	0.0
10	lost target	-0.2
11	dropping an object at destination	0.0
12	grabbing an object	0.0
13	failed grabbing or dropping	-1
14	found object 1	0.3
17	found the destination	0.0
16	ignoring an object	-0.1

5.3 Performance evaluation in measure μ

Both the state transition cost (Π) matrix and the characteristic $\mathbf{X}$-vector are needed for performance evaluation in language measure μ. The states of plant model G and their associated χ values are listed Table 5.3 and the Π–matrix is listed in Table 5.4.

A χ-value of 0.4 is assigned to state 8 where the robot grabs an object successfully. The state where the robot successfully drops an object at the destination, which also represents the end of the current mission, is assigned a χ-value of 0.8. The two colored object are given different level of importance.

Table 5.4. The Π-matrix of the unsupervised plant model G

0.9500	0	0	0	0	0	0	0	0	0	0	0	0	0	0	0	0
0.4087	0.0254	0	0	0	0	0	0	0.4087	0	0	0	0	0.0288	0.0785	0	0
0	0	0	0.4396	0	0	0	0	0	0	0	0	0	0	0	0	0.5104
0	0	0	0	0	0	0	0	0	0	0	0	0	0	0	0	0
0	0	0	0	0.3760	0.3760	0	0	0	0.1979	0	0	0	0	0	0	0
0	0	0	0	0	0	0	0	0	0	0.95	0	0	0	0	0	0
0	0	0	0	0	0	0	0	0	0	0	0.95	0	0	0	0	0
0.9500	0	0	0	0	0	0	0	0	0	0	0	0	0	0	0	0
0.9500	0	0	0	0	0	0	0	0	0	0	0	0	0	0	0	0
0.4750	0	0	0	0	0	0	0	0.4750	0	0	0	0	0	0	0	0
0.9500	0	0	0	0	0	0	0	0	0	0	0	0	0	0	0	0
0	0	0	0	0	0.95	0	0	0	0	0	0	0	0	0	0	0
0	0	0	0	0	0	0.95	0	0	0	0	0	0	0	0	0	0
0	0	0	0	0	0	0	0	0	0	0	0	0	0	0	0	0
0	0	0	0.4495	0	0	0	0	0	0	0	0	0	0	0	0	0.5005
0	0	0	0.5200	0	0	0	0	0	0	0	0	0	0	0	0	0.4300
0.3484	0	0	0	0	0	0	0	0	0	0	0	0	0	0	0	0.6016

For example, the credit for discovering a pink object, is $\chi = 0.3$ and that for a green object is $\chi = 0.1$. Losing and ignoring an object is given a penalty of -0.2 and -0.1, respectively. Following Chapter 1, the performance comparisons between the unsupervised and supervised systems for both robot experiments and robot simulation are given by:

$$\mu_{\text{Experiment}}(L(G)) = 0.1640, \qquad \mu_{\text{Experiment}}(L(S/G)) = 0.1987$$
$$\mu_{\text{Simulation}}(L(G)) = 0.0799, \qquad \mu_{\text{Simulation}}(L(S/G)) = 0.1349$$

The theoretical results based on robot experiments and robot simulation are consistent in view of the performance of the supervised (i.e., closed loop) system S/G relative to that of the unsupervised (i.e., open loop) plant G as shown in Figure 5.19. In missions of object collection lasting about five hours, the robot under DES control performs much better both in terms of language-measure-based theoretical performance and actual performance in number of successful missions. As expected, whenever the robot finds an object, the DES controller always allows the robot to approach the object and grab it. In contrast, without the DES control, the robot may ignore the object and keep on searching. It is shown in Chapter 6 that the performance of a DES controller may change when operating in coordination with other robots. This happens because the event cost matrix $\tilde{\Pi}$ changes in the altered environment resulting from participation of other robot(s).

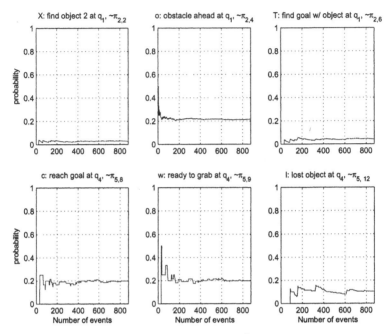

Fig. 5.17. Convergence of selected non-zero $\tilde{\Pi}$ elements in experiment

5.4 Conclusions

This chapter presents the details of a discrete-event behavior-based design of a robot system in the real-time environment as well as its simulator. The robot simulator demonstrates its high fidelity with respect to the actual robot; hence, the simulator can be used for further investigation of more complex scenarios. For an experiment scenario, a discrete-event model G of the unsupervised plant is formulated and a DES controller S is designed. The event cost matrix $\tilde{\Pi}$ is identified according to the procedure presented in Chapter 1 in both robot experiment and simulation. The characteristic vector $\mathbf{X}$ is assigned according to the mission objective of the experiment scenario. Through both theoretical language measure computation and empirical cumulative performance, it is shown that the supervised robot yields better performance than the unsupervised robot; the language measure μ is proven to be a useful quantitative measure for performance evaluation of a discrete-event supervisor.

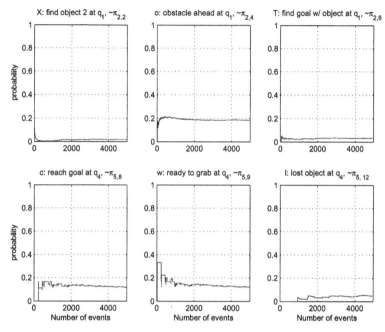

Fig. 5.18. Convergence of selected non-zero $\tilde{\Pi}$ elements in simulation

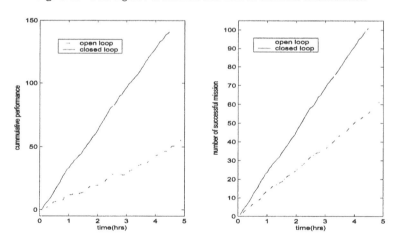

Fig. 5.19. Cumulative performance comparison in simulation

References

1. J. R. Andrews and N. Hogan, *Impedance control as a framework for implementing obstacle avoidance in a manipulator*, Control of Manufacturing Processes and Robotic Systems (D. E. Hardt and W. Book, eds.), ASME, Boston, 1983, pp. 243–251.

2. J. Borenstein and Y. Koren, *Histogramic in-motion mapping for mobile robot obstacle avoidance*, IEEE Transaction on Robotics and Automation (1991), 535–539.

3. _____, *The vector field histogram - fast obstacle avoidance for mobile robots*, IEEE Transaction on Robotics and Automation **7** (1991), no. 3, 278–288.

4. R. W. Brockett, R. S. Millmann, and J. H. Sussmann, *Differential geometric control theory*, ch. Asymptotic stability and feedback stabilization, pp. 181–191, Birkhauser, Boston, 1983.

5. R.A. Brooks, *A robust layered control system for a mobile robot*, IEEE Journal of Robotics and Automation **2** (1986), no. 1, 14–23.

6. _____, *Intelligence without representation*, Artifical Intelligence **47** (1991), 139–160.

7. Y.U. Cao, A.S. Fukunaga, and A.B. Kahng, *Cooperative mobile robotics: Antecedents and directions*, Autonomous Robots **4** (1997), 1–23.

8. J. T. Feddema, R. D. Robinett, and B. J. Driesen, *Designing stable finite state machine behaviours using phase plane analysis and variable structure control*, May 1998, pp. 1134–1141.

9. B. P. Gerkey, R. T. Vaughan, and A. Howard, *The player/stage project: tools for multi-robot and distributed sensor systems*, June 2003, pp. 317–323.

10. O. Khatib, *Real-time obstacle avoidance for manipulators and mobile robots*, Int. Journal of Robotics Research **5** (1986), no. 1, 90–98.

11. Y. Koren and J. Borenstein, *Potential field methods and their inherent limitation for mobile robot navigation*, Proceedings of IEEE International Conference on Robotics and Automation, 1991, pp. 1398–1404.

12. J. Kosecka, *A framework for modeling and verifying visually giuded agents: Design, analysis and experiments*, Ph.D. thesis, University of Pennsylvania, March 1996.

13. Y. Ma, *A differential geometric approach to computer vision and its applications in control*, Ph.D. thesis, EECS, UC Berkeley, August 2000.

14. P. Maes and R. A. Brook, *Learning to coordinate behaviors*, AAAI (1990), 796–802.

15. P. V. Overschee and B. D. Moor, *Identification for linear system theory - implementation - applications*, Kluwer Academic Publishers, 1996.

16. P.J. Ramadge and W.M. Wonham, *Supervisory control of a class of discrete event processes*, SIAM J. Control and Optimization **25** (1987), no. 1, 206–230.

17. I. Ulrich and J. Borenstein, *Vfh+: reliable obstacle avoidance for fast mobile robots*, Proceedings of the 1998 IEEE International Conference on Robotics and Automation, May 1998, pp. 1572–1577.

6

Optimal Control of Robot Behavior Using Language Measure

Xi Wang[1], Asok Ray[2], Peter Lee[3], and Jinbo Fu[4]

[1] The Pennsylvania State University xxw117@psu.edu
[2] The Pennsylvania State University axr2@psu.edu
[3] The Pennsylvania State University cfl106@psu.edu
[4] The Pennsylvania State University jinbofu@adelphia.net

Summary. This chapter presents optimal discrete-event supervisory control of robot behavior in terms of the language measure μ, presented in Chapter 1. In the
discrete-event setting, a robot's behavior is modelled as a regular language that can
be realized by deterministic finite state automata (DFSA). The controlled sublanguage of a DFSA plant model could be different under different supervisors that
are constrained to satisfy different specifications [6]. Such a partially ordered set of
sublanguages requires a quantitative measure for total ordering of their respective
performance. The language measure [10] [8] serves as a common quantitative tool
to compare the performance of different supervisors and is assigned an event cost
matrix, known as the $\widetilde{\Pi}$-matrix and a state characteristic vector, **X**-vector. Event
costs (i.e., elements of the $\widetilde{\Pi}$-matrix) are based on the plant states, where they are
generated; on the other hand, the **X**-vector is chosen based on the designer's perception of the individual state's impact on the system performance. The elements of
the $\widetilde{\Pi}$-matrix are conceptually similar to the probabilities of the respective events
conditioned on specific states; these parameters can be identified either from experimental data or from the results of extensive simulation, as they are dependent on
physical phenomena related to the plant behavior. Since the plant behavior is often
slowly time-varying, there is a need for on-line parameter identification to generate
up-to-date values of the $\widetilde{\Pi}$-matrix within allowable bounds of errors. The results of
simulation experiments on a robotic test bed are presented to demonstrate efficacy
of the proposed optimal control policy.

Key words: Discrete Event Supervisory Control, Robot Behavior Control,
Computer Vision, Language Measure

6.1 Introduction

In the discrete-event setting, the behavior of physical plants, such as a robot,
is often modelled as regular languages that can be realized by deterministic
finite state automata (DFSA) in the discrete-event setting [6]. This chapter introduces and validates a novel approach for robot behavior selection
based on discrete event supervisory control in terms of the language measure μ [10] [7] [8]. The controlled sublanguage of a DFSA plant model could
be different under different supervisors that are constrained to satisfy different specifications. Such a partially ordered set of sublanguages requires a
quantitative measure for total ordering of their respective performance. The
language measure serves as a common quantitative tool to compare the performance of different supervisors and is assigned an event cost matrix and a
state characteristic vector.

This chapter presents discrete-event optimal supervisory control of mobile
robot operations, where the control policy is synthesized by maximizing the
language measure of robot behavior. This novel approach is called μ-optimal
in the sequel. In contrast to the Q-learning reinforcement [11] [12] that has
been widely used in robotics, computational complexity of μ-optimal control is
polynomial in the number of states of the deterministic finite state automaton

(DFSA) model of unsupervised robot behavior. The simulation results and the experimental data on a robotic test bed are presented for typical scenarios to demonstrate efficacy of the μ-optimal control policy.

The chapter is organized in seven sections including the present section. The language measure is briefly reviewed in Section 6.2 including introduction of the notations. Section 6.3 presents the parameter estimation algorithm and the associated stopping rule. Section 6.4 proposes a discrete event supervisory (DES) control synthesis strategy. Section 6.5 validates the synthesis algorithm through simulation and actual experiments on a mobile robotic test bed. Section 6.6 succinctly compares the proposed language-based supervisory control with the Q-learning method [11] that is widely used for reinforcement in behavior-based robotics. The chapter is summarized and concluded in Section 6.7.

6.2 Quantitative Measure of Regular Languages for Supervisory Control

This section introduces the notion of language measure [10] [7] [8] that assigns an event cost matrix, denoted as the $\widetilde{\Pi}$-matrix, and a state characteristic vector, denoted as the $\mathbf{X}$-vector. Event costs (i.e., elements of the $\widetilde{\Pi}$-matrix) are based on plant states, where they are generated, are physical phenomena dependent on the plant behavior, and are conceptually similar to the conditional probabilities of the respective events. On the other hand, the characteristic vector, denoted as the $\mathbf{X}$-vector, is chosen based on the designer's perception of the individual state's impact on the system performance. In the performance evaluation of both the unsupervised and supervised plant behavior, the critical parameter is the event cost $\widetilde{\Pi}$-matrix. Since the plant behavior is often slowly time-varying, there is a need for on-line parameter identification to generate up-to-date values of the $\widetilde{\Pi}$-matrix within allowable bounds of errors.

Let $G_i \equiv \langle Q, \Sigma, \delta, q_i, Q_m \rangle$ be a trim (i.e., accessible and co-accessible) finite-state automaton model that represents the discrete-event dynamics of a physical plant where $Q = \{q_k : k \in \mathcal{I}_Q\}$, where $\mathcal{I}_Q \equiv \{1, 2, \cdots, n\}$, is the set of states with q_i, where $i \in \mathcal{I}_Q$, being the initial state; $\Sigma = \{\sigma_k : k \in \mathcal{I}_\Sigma\}$, where $\mathcal{I}_\Sigma \equiv \{1, 2, \cdots, \ell\}$, is the alphabet of events; $\delta : Q \times \Sigma \to Q$ is the (possibly partial) function of state transitions; and $Q_m \equiv \{q_{m_1}, q_{m_2}, \cdots, q_{m_l}\} \subseteq Q$ is the set of marked (i.e., accepted) states $q_{m_k} = q_j$ for some $j \in \mathcal{I}_Q$.

Let Σ^* be the Kleene closure of Σ, i.e., the set of all finite-length strings made of the events belonging to Σ as well as the empty string ϵ that is viewed as the identity of the monoid Σ^* under the operation of string concatenation, i.e., $\epsilon s = s = s\epsilon$. The extension $\hat{\delta} : Q \times \Sigma^* \to Q$ is defined recursively in the usual sense [6].

Definition 6.2.1 *The language $L(G_i)$ generated by a DFSA G initialized at the state $q_i \in Q$ is defined as:*

$$L(G_i) = \{s \in \Sigma^* \mid \hat{\delta}(q_i, s) \in Q\} \tag{6.1}$$

The language $L_m(G_i)$ marked by the DFSA G initialized at the state $q_i \in Q$ is defined as:

$$L_m(G_i) = \{s \in \Sigma^* \mid \hat{\delta}(q_i, s) \in Q_m\} \tag{6.2}$$

Definition 6.2.2 *For every $q_j \in Q$, let $L(q_i, q_j)$ denote the set of all strings that, starting from the state q_i, terminate at the state q_j, i.e.,*

$$L(q_i, q_j) = \{s \in \Sigma^* \mid \hat{\delta}(q_i, s) = q_j \in Q\} \tag{6.3}$$

The set Q_m of marked states is partitioned into Q_m^+ and Q_m^-, i.e., $Q_m = Q_m^+ \cup Q_m^-$ and $Q_m^+ \cap Q_m^- = \emptyset$, where Q_m^+ contains all *good* marked states that we desire to reach, and Q_m^- contains all *bad* marked states that we want to avoid, although it may not always be possible to completely avoid the *bad* states while attempting to reach the *good* states. To characterize this, each marked state is assigned a real value based on the designer's perception of its impact on the system performance.

Definition 6.2.3 *The characteristic function $\chi : Q \rightarrow [-1, 1]$ that assigns a signed real weight to state-based sublanguages $L(q_i, q)$ is defined as:*

$$\forall q \in Q, \quad \chi(q) \in \begin{cases} [-1, 0), & q \in Q_m^- \\ \{0\}, & q \notin Q_m \\ (0, 1], & q \in Q_m^+ \end{cases} \tag{6.4}$$

The state weighting vector, denoted by $\mathbf{X} = [\chi_1\, \chi_2\, \cdots\, \chi_n]^T$, where $\chi_j \equiv \chi(q_j)$ $\forall k$, is called the $\mathbf{X}$-vector. The j-th element χ_j of $\mathbf{X}$-vector is the weight assigned to the corresponding terminal state q_j.

In general, the marked language $L_m(G_i)$ consists of both good and bad event strings that, starting from the initial state q_i, lead to Q_m^+ and Q_m^- respectively. Any event string belonging to the language $L^0 = L(G_i) - L_m(G_i)$ leads to one of the non-marked states belonging to $Q - Q_m$ and L^0 does not contain any one of the good or bad strings. Based on the equivalence classes defined in the Myhill-Nerode Theorem, the regular languages $L(G_i)$ and $L_m(G_i)$ can be expressed as:

$$L(G_i) = \bigcup_{q_k \in Q} L(q_i, q_k) = \bigcup_{k=1}^{n} L(q_i, q_k) \tag{6.5}$$

$$L_m(G_i) = \bigcup_{q_k \in Q_m} L(q_i, q_k) = L_m^+ \cup L_m^- \tag{6.6}$$

where the sublanguage $L(q_i, q_k) \subseteq G_i$ having the initial state q_i is uniquely labeled by the terminal state $q_k, k \in \mathcal{I}$ and $L(q_i, q_j) \cap L(q_i, q_k) = \emptyset \; \forall j \neq k$; and $L_m^+ \equiv \bigcup_{q \in Q_m^+} L(q_i, q)$ and $L_m^- \equiv \bigcup_{q \in Q_m^-} L(q_i, q)$ are good and bad sublanguages of $L_m(G_i)$, respectively. Then, $L^0 = \bigcup_{q \notin Q_m} L(q_i, q)$ and $L(G_i) = L^0 \cup L_m^+ \cup L_m^-$.

Now we construct a signed real measure $\mu : 2^{L(G_i)} \to \mathbf{R} \equiv (-\infty, +\infty)$ on the σ-algebra $K = 2^{L(G_i)}$. Interested readers are referred to [10] for the details of measure-theoretic definitions and results. With the choice of this σ-algebra, every singleton set made of an event string $\omega \in L(G_i)$ is a measurable set, which qualifies itself to have a numerical quantity based on the above state-based decomposition of $L(G_i)$ into L^0(null), L^+(positive), and L^-(negative) sublanguages.

Conceptually similar to the conditional transition probability, each event is assigned a state-dependent cost.

Definition 6.2.4 *The event cost of the DFSA G_i is defined as a (possibly partial) function $\tilde{\pi} : Q \times \Sigma^* \to [0, 1)$ such that $\forall q_i \in Q$, $\forall \sigma_j \in \Sigma$, $\forall s \in \Sigma^*$,*

(1) $\tilde{\pi}[q_i, \sigma_j] \equiv \tilde{\pi}_{ij} \in [0, 1)$; $\sum_j \tilde{\pi}_{ij} < 1$;
(2) $\tilde{\pi}[q_i, \sigma_j] = 0$ if $\delta(q_i, \sigma_j)$ is undefined; $\tilde{\pi}[q_i, \epsilon] = 1$;
(3) $\tilde{\pi}[q_i, \sigma_j s] = \tilde{\pi}[q_i, \sigma_j]\, \tilde{\pi}[\delta(q_i, \sigma_j), s]$.

Definition 6.2.5 *The state transition cost of the DFSA G_i is defined as a function $\pi : Q \times Q \to [0, 1)$ such that $\forall q_i, q_j \in Q$, $\pi[q_i, q_j] = \sum_{\sigma \in \Sigma : \delta(q_i, \sigma) = q_j} \tilde{\pi}[q_i, \sigma] \equiv \pi_{ij}$ and $\pi_{ij} = 0$ if $\{\sigma \in \Sigma : \delta(q_i, \sigma) = q_j\} = \emptyset$. The $n \times n$ state transition cost $\mathbf{\Pi}$-matrix is defined as:*

$$\mathbf{\Pi} = \begin{bmatrix} \pi_{11} & \pi_{12} & \cdots & \pi_{1n} \\ \pi_{21} & \pi_{22} & \cdots & \pi_{2n} \\ \vdots & \vdots & \ddots & \vdots \\ \pi_{n1} & \pi_{n2} & \cdots & \pi_{nn} \end{bmatrix}$$

Definition 6.2.6 *The real signed measure μ of every singleton string set $\Omega = \{\omega\} \in 2^{L(G_i)}$ where $\omega \in L(q_i, q)$ is defined as $\mu(\Omega) \equiv \tilde{\pi}[q_i, \omega]\chi(q)$. It follows that the signed real measure of the sublanguage $L(q_i, q) \subseteq L(G_i)$ is defined as*

$$\mu(L(q_i, q)) = \left(\sum_{\omega \in L(q_i, q)} \tilde{\pi}[q_i, \omega] \right) \chi(q) \tag{6.7}$$

And the signed real measure of the language of a DFSA G_i initialized at a state $q_i \in Q$, is defined as:

$$\mu_i \equiv \mu(L(G_i)) = \sum_{q \in Q} \mu(L(q_i, q)) \tag{6.8}$$

The language measure vector, denoted as $\boldsymbol{\mu} = [\mu_1 \ \mu_2 \ \cdots \ \mu_n]^T$, is called the μ-vector.

It is shown in [10] [8] that the signed real measure μ_i can be written as:

$$\mu_i = \sum_j \pi_{ij}\mu_j + \chi_i \tag{6.9}$$

In vector form, $\boldsymbol{\mu} = \boldsymbol{\Pi}\boldsymbol{\mu} + \mathbf{X}$ whose solution is given by

$$\boldsymbol{\mu} = (\mathbf{I} - \boldsymbol{\Pi})^{-1}\mathbf{X} \tag{6.10}$$

The inverse in Eq. (6.10) exists because Π is a contraction operator [10] [8].

6.2.1 Probabilistic Interpretation

Since the plant model is an inexact representation of the physical plant, there exist unmodeled dynamics to be accounted for. This can manifest itself either as unmodeled events that may occur at each state or as unaccounted states in the model. Let Σ_k^u denote the set of all unmodeled events at state k of the DFSA $G_i \equiv \langle Q, \Sigma, \delta, q_i, Q_m \rangle$. Let us create a new unmarked absorbing state q_{n+1}, called the dump state [6], and extend the transition function δ to $\delta_e : (Q \cup \{q_{n+1}\}) \times (\Sigma \cup (\cup_k \Sigma_k^u)) \to (Q \cup \{q_{n+1}\})$ as follows:

$$\delta_e(q_k, \sigma) = \begin{cases} \delta(q_k, \sigma) & \text{if } q_k \in Q \text{ and } \sigma \in \Sigma \\ q_{n+1} & \text{if } q_k \in Q \text{ and } \sigma \in \Sigma_k^u \\ q_{n+1} & \text{if } k = n+1 \text{ and } \sigma \in \Sigma \cup \Sigma_k^u \end{cases}$$

Therefore the residue $\theta_k = 1 - \sum_j \tilde{\pi}_{kj}$ denotes the probability of the set of unmodeled events Σ_k^u conditioned on the state q_k. The $\boldsymbol{\Pi}$-matrix is accordingly augmented to obtain a stochastic matrix $\boldsymbol{\Pi}^{aug}$ as follows:

$$\boldsymbol{\Pi}^{aug} = \begin{bmatrix} \pi_{11} & \pi_{12} & \cdots & \pi_{1n} & \theta_1 \\ \pi_{21} & \pi_{22} & \cdots & \pi_{2n} & \theta_2 \\ \vdots & \vdots & \ddots & \vdots & \vdots \\ \pi_{n1} & \pi_{n2} & \cdots & \pi_{nn} & \theta_n \\ 0 & 0 & \cdots & 0 & 1 \end{bmatrix}$$

Since the dump state q_{n+1} is not marked, its characteristic value $\chi(q_{n+1}) = 0$. The characteristic vector then augments to $\mathbf{X}^{aug} \equiv [\mathbf{X}^T \ 0]^T$. With these extensions the language measure vector $\boldsymbol{\mu}^{aug} = [\mu_1 \ \mu_2 \ \cdots \ \mu_n \ \mu_{n+1}]^T = [\boldsymbol{\mu} \ \mu_{n+1}]^T$ of the augmented DFSA $G_i^{aug} \equiv (Q \cup \{q_{n+1}\}, \Sigma \cup (\cup_k \Sigma_k^u), \delta_e, q_i, Q_m)$ can be expressed as:

$$\begin{pmatrix} \boldsymbol{\mu} \\ \mu_{n+1} \end{pmatrix} = \begin{pmatrix} \boldsymbol{\Pi}\boldsymbol{\mu} + \mu_{n+1} \left[\theta_1 \cdots \theta_n \right]^T \\ \mu_{n+1} \end{pmatrix} + \begin{pmatrix} \mathbf{X} \\ 0 \end{pmatrix} \tag{6.11}$$

Since $\chi(q_{n+1}) = 0$ and all transitions from the absorbing state q_{n+1} lead to itself, $\mu_{n+1} = \mu(L_m(G_{n+1})) = 0$. Hence, Equation (6.11) reduces to that for the original plant DFSA G_i. Thus, the event cost can be interpreted as the conditional probability, where the residue $\theta_k = (1 - \sum_j \tilde{\pi}_{kj}) \in (0, 1]$ accounts for the probability of all unmodeled events emanating from the state q_k.

6.3 Estimation of Language Measure Parameters

This section presents a recursive algorithm for identification of the language measure parameters (i.e., elements of the event cost matrix $\widetilde{\boldsymbol{\Pi}}$) (see Definition 6.2.4) which, in turn, allows computation of the state transition cost matrix $\boldsymbol{\Pi}$ (see Definition 6.2.5) and the language measure $\boldsymbol{\mu}$-vector (see Definition 6.2.6). It is assumed that the underlying physical process evolves at two different time scales. In the fast-time scale, i.e., over a short time period, the system is assumed to be an ergodic, discrete Markov process. In the slowly-varying time scale, i.e., over a long period, the system (possibly) behaves as a non-stationary stochastic process. For such a slowly-varying non-stationary process, it might be necessary to redesign the supervisory control policy in real time. In that case, the $\widetilde{\boldsymbol{\Pi}}$-matrix parameters should be periodically updated.

6.3.1 A Recursive Parameter Estimation Scheme

Let p_{ij} be the transition probability of the event σ_j at the state q_i, i.e.,

$$p_{ij} = \begin{cases} P[\sigma_j | q_i], & \text{if } \exists q \in Q, \ s.t. \ q = \delta(q_i, \sigma_j) \\ 0, & \text{otherwise} \end{cases} \tag{6.12}$$

and its estimate be denoted by the parameter $\hat{p}_{ij}$ that is to be identified from the ensemble of simulation and/or experimental data.

Let a strictly increasing sequence of time epochs of consecutive event occurrence be denoted as:

$$\mathcal{T} \equiv \{t_k : k \in \mathbf{N}_0\} \tag{6.13}$$

where $\mathbf{N}_0$ is the set of non-negative integers. Let the indicator $\psi : \mathbf{N}_0 \times \mathcal{I}_Q \times \mathcal{I}_\Sigma \rightarrow \{0, 1\}$ represent the incident of occurrence of an event. For example, if the DFSA was in state q_i at time epoch t_{k-1}, then

$$\psi_{ij}(k) = \begin{cases} 1, & \text{if } \sigma_j \text{ occurs at time epoch } t_k \in \mathcal{T} \\ 0, & \text{otherwise} \end{cases} \tag{6.14}$$

Consequently, the number of occurrences of any event in the alphabet Σ is represented by $\Psi : \mathbf{N}_0 \times \mathcal{I}_Q \to \{0, 1\}$. For example, if the DFSA was in state q_i at the time epoch t_{k-1}, then

$$\Psi_i(k) = \sum_{j \in \mathcal{I}_\Sigma} \psi_{ij}(k) \tag{6.15}$$

Let $n : \mathbf{N}_0 \times \mathcal{I}_Q \times \mathcal{I}_\Sigma \to \mathbf{N}_0$ represent the cumulative number of occurrences of an event at a state up to a given time epoch. That is, $n_{ij}(k)$ denotes the number of occurrences of the event σ_j at the state q_i up to the time epoch $t_k \in \mathcal{T}$. Similarly, let $N : \mathbf{N}_0 \times \mathcal{I}_Q \to \mathbf{N}_0$ represent the cumulative number of occurrences of any event in the alphabet Σ at a state up to a given time epoch. Consequently,

$$N_i(k) = \sum_{j \in \mathcal{I}_\Sigma} n_{ij}(k) \tag{6.16}$$

A frequency estimator, $\hat{p}_{ij}(k)$, for probability $p_{ij}(k)$ of the event σ_j occurring at the state q_i at the time epoch t_k, is obtained as:

$$\hat{p}_{ij}(k) = \frac{n_{ij}(k)}{N_i(k)}$$
$$\lim_{k \to \infty} \hat{p}_{ij}(k) = p_{ij} \tag{6.17}$$

A recursive algorithm of learning p_{ij} is formulated as a stochastic approximation scheme, starting at the time epoch t_0 with the initial conditions: $\hat{p}_{ij}(0) = 0$ and $n_{ij}(0) = 0$ for all $i \in \mathcal{I}_Q, j \in \mathcal{I}_\Sigma$; and $\Psi_i(0) = 0$ for all $i \in \mathcal{I}_Q$. Starting at $k = 0$, the recursive algorithm runs for $\{t_k : k \geq 1\}$. For example, upon occurrence of an event σ_j at a state q_i, the algorithm is recursively incremented as:

$$n_{ij}(k) = n_{ij}(k-1) + \psi_{ij}(k)$$
$$N_i(k) = N_i(k-1) + \Psi_i(k) \tag{6.18}$$

Next it is demonstrated how the estimates of the language parameters (i.e., the elements of event cost matrix $\tilde{\Pi}$) are determined from the probability estimates. As stated earlier in Section 6.2.1, the set of unmodelled events at state q_i, denoted by $\Sigma_i^u \; \forall i \in \mathcal{I}_Q$, accounts for the row-sum inequality: $\sum_j \tilde{\pi}_{ij} < 1$ (see Definition 6.2.4). Then, $P[\Sigma_i^u] = \theta_i \in (0, 1)$ and $\sum_i \tilde{\pi}_{ij} = 1 - \theta_i$. An estimate of the $(i, j)^{th}$ element of the $\tilde{\Pi}$-matrix, denoted by $\hat{\tilde{\pi}}_{ij}$, is approximated as:

$$\hat{\tilde{\pi}}_{ij}(k) = \hat{p}_{ij}(k)(1 - \theta_i) \quad \forall j \in \mathcal{I}_\Sigma \tag{6.19}$$

Additional experiments on a more detailed automaton model would be necessary to identify the parameters $\theta_i \; \forall i \in \mathcal{I}_Q$. Given that $\theta_i \ll 1$, the problem of conducting additional experimentation can be circumvented by the following approximation:

A single parameter $\theta \approx \theta_i \ \forall i \in \mathcal{I}_Q, \ i \in \mathcal{I}_Q$, such that $0 < \theta \ll 1$, could be selected for convenience of implementation. From the numerical perspective, this option is meaningful because it sets an upper bound on the language measure based on the fact that the sup-norm $\|\mu\|_\infty \leq \theta^{-1}$. Note that each row sum in the $\widetilde{\Pi}$-matrix being strictly less than 1, i.e., $\sum_j \tilde{\pi}_{ij} < 1$, is a sufficient condition for finiteness of the language measure.

Theoretically, $\tilde{\pi}_{ij}$ is the asymptotic value of the estimated probabilities $\hat{\tilde{\pi}}_{ij}(k)$ as if the event σ_j occurs infinitely many times at the state q_i. However, dealing with finite amount of data, the objective is to obtain a *good* estimate $\hat{p}_{ij}$ of p_{ij} from independent Bernoulli trials of generating events. Critical issues in dealing with finite amount of data are: (i) how much data are needed; and (ii) when to stop if adequate data are available. The next section 6.3.2 addresses these issues.

6.3.2 A Stopping Rule for Recursive Learning

A stopping rule is proposed to find a lower bound on the number of experiments to be conducted for the $\widetilde{\Pi}$-matrix parameter identification. This section presents such a stopping rule for an inference approximation having a specified absolute error bound ε with a probability λ. The objective is to achieve a trade-off between the number of experimental observations and the estimation accuracy. A robust stopping rule is presented below.

A bound on the required number of samples is estimated using the Gaussian structure for binomial distribution that is an approximation of the sum of a large number of independent and identically distributed (i.i.d.) Bernoulli trials of $\hat{\tilde{\pi}}_{ij}(t)$. The central limit theorem yields $\hat{\tilde{\pi}}_{ij} \sim \mathcal{N}(\tilde{\pi}_{ij}, \frac{\tilde{\pi}_{ij}(1-\tilde{\pi}_{ij})}{N})$, where $\mathcal{N}$ indicates normal (or Gaussian) distribution with $E[\hat{\tilde{\pi}}_{ij}] \approx \tilde{\pi}_{ij}$ and $\mathrm{Var}[\hat{\tilde{\pi}}_{ij}] \equiv \sigma^2 \approx \frac{\tilde{\pi}_{ij}(1-\tilde{\pi}_{ij})}{N}$, provided that the number of samples N is sufficiently large. Let $\Delta = \hat{\tilde{\pi}}_{ij} - \tilde{\pi}_{ij}$, then $\frac{\Delta}{\sigma} \sim \mathcal{N}(0,1)$. Given $0 < \varepsilon \ll 1$ and $0 < \lambda \ll 1$, the problem is to find a bound N_b on the number N of experiments such that $P\{|\Delta| \geq \varepsilon\} \leq \lambda$. Equivalently,

$$P\left\{\frac{|\Delta|}{\sigma} \geq \frac{\varepsilon}{\sigma}\right\} \leq \lambda \tag{6.20}$$

that yields a bound N_b on N as:

$$N_b \geq \left(\frac{\xi^{-1}(\lambda)}{\varepsilon}\right)^2 \tilde{\pi}_{ij}(1 - \tilde{\pi}_{ij}) \tag{6.21}$$

where $\xi(x) \equiv 1 - \sqrt{\frac{2}{\pi}} \int_0^x e^{-\frac{t^2}{2}} dt$. Since the parameter $\tilde{\pi}_{ij}$ is unknown, one may use the fact that $\tilde{\pi}_{ij}(1 - \tilde{\pi}_{ij}) \leq 0.25$ for every $\tilde{\pi}_{ij} \in [0,1]$ to (conservatively) obtain a bound on N only in terms of the specified parameters ε and λ as:

$$N_b \geq \left(\frac{\xi^{-1}(\lambda)}{2\varepsilon} \right)^2 \tag{6.22}$$

The above estimate of the bound on the required number of samples is less conservative than that obtained from the Chernoff bound and is significantly less conservative than that obtained from Chebyshev bound [5] that does not require the assumption of any specific distribution of Δ except for finiteness of the rth $(r = 2)$ moment.

6.4 Discrete Event Supervisory Control Synthesis in μ-measure

In the conventional discrete event supervisory (DES) control synthesis [6], the qualitative measure of maximum permissiveness plays an important role. For example, under full state observation, if a specification language K is not controllable with respect to the plant automaton G and the set Σ_u of uncontrollable events, then a supremal controllable sublanguage $SupC(K) \subseteq K$ yields maximal permissiveness. However, increased permissiveness of the controlled language $L(S/G)$ may not generate better plant performance from the perspectives of mission accomplishment. This section relies on the language measure μ to quantitatively synthesize a Discrete Event Supervisory (DES) control policy. The objective is to design a supervisor such that the controlled plant automaton S/G maximizes the performance that is chosen as the measure μ of the controlled plant language $L(S/G)$. The pertinent assumptions for the DES control synthesis are delineated below.

A1 (Cost redistribution) The probabilities of occurrence of *controllable* events in a controlled sublanguage $L(S/G) \subseteq L(G)$ are proportional to those in $L(G)$. For all $q \in Q_S$, where Q_S is the state space of the supervisor automaton S, and $\sigma \in \Sigma_S(q)$, where $\Sigma_S(q)$ is the set of events defined at $q \in Q_S$.

$$\tilde{\pi}_S[q, \sigma] = \frac{\tilde{\pi}_G[q, \sigma]}{\sum_{\sigma \in \Sigma_S(q)} \tilde{\pi}_G[q, \sigma]} \tag{6.23}$$

A2 (Event controllability) Any transition $\delta(q, \sigma)$, defined in the plant automaton G such that $\sigma \in \Sigma_{uc}(G)$ and $q \in Q$, is kept enabled in a supervisor S.

Under assumption A1, the sum of event costs defined at the state q of a supervisor S is equal to that of state q of the plant G, i.e.,

$$\sum_{\sigma \in \Sigma_S(q)} \tilde{\pi}_S[q, \sigma] = \sum_{\sigma \in \Sigma_G(q)} \tilde{\pi}_G[q, \sigma] \tag{6.24}$$

Lemma 6.4.1 (Finiteness) *By disabling controllable events in a plant automaton G, there is only a finite number of controllers $S^i, i \in \mathcal{I}_c$, where $\mathcal{I}_c$ is the set of controllers with cardinality $|\mathcal{I}_c| = n_c$, such that for every $i \in \mathcal{I}_c$, $L(S^i) = L(S^i/G) \subseteq L(G)$.*

Proof. Under assumption A2, it suffices to show that the number of all possible permutations of disabling controllable events defined on all states in G is finite. The worst case is that: (1) for every state $q \in Q$ and every controllable event $\sigma \in \Sigma_c$, the transition $\delta(q, \sigma)$ is defined and; (2) every state q in G does not depend on any other state to be accessible from initial state q_0. Then, the number of all possible transitions n_t is given by:

$$n_t \leq |Q| \times |\Sigma_c| = nm \tag{6.25}$$

where $|Q| = n$ is the number of states and $|\Sigma_c| = m$ is the number of controllable events in G. And the number of all possible supervisors is given by

$$n_c \leq \binom{n_t}{0} + \binom{n_t}{1} + \binom{n_t}{2} + \cdots + \binom{n_t}{n_t} = \sum_{i=0}^{n_t} \binom{n_t}{i} < \infty \tag{6.26}$$

Lemma 6.4.1 shows that there are finitely many supervisors whose generating language is a subset of $L(G)$ given the fact that both the state space and event alphabet are finite. Next we present pertinent results in the form of two theorems. Theorem 6.4.1 states that out of all possible supervisors constructed from G, there exists a supervisor that maximizes the language measure μ with respect to $(\mathbf{\Pi}, \mathbf{X})$. Theorem 6.4.2 describes a general transition structure of the μ-optimal supervisor S^*. In particular, at every state in $q \in Q_c(S^*)$ in S^*, there is one and only one controllable event defined.

Theorem 6.4.1 (Existence) *[2] Given a DFSA plant $G = (Q, \Sigma, \delta, q_0, Q_m)$, $\mathbf{\Pi}$-matrix, and $\mathbf{X}$-vector, there exist an optimal supervisor S^* such that $\mu(L(S^*/G)) = \max_{i \in \mathcal{I}_c} \mu(L(S_i/G))$.*

Theorem 6.4.2 *Given a DFSA plant $G = (Q, \Sigma, \delta, q_0, Q_m)$, $\mathbf{\Pi}$-matrix, and $\mathbf{X}$-vector, the event set $\Sigma_c(G, q)$ and the plant set $Q_c(G)$ are defined as follows:*

$$\Sigma_c(G, q) = \{\sigma \in \Sigma_c \mid \delta(q, \sigma) \text{ is defined in } G\} \tag{6.27}$$

$$Q_c(G) = \{q \in Q \mid \Sigma_c(G, q) \neq \emptyset\} \tag{6.28}$$

For every state $q \in Q_c(G)$, there is one and only one controllable event left enabled in the μ-optimal supervisor S^*, i.e.,

$$\forall q \in Q_c(S^*), \qquad |\Sigma_c(S^*, q)| = 1 \tag{6.29}$$

Proof. Let us assume that there exists a supervisor S^ι such that $\mu(L(S^\iota/G)) > \mu(L(S^*/G))$ and $|\Sigma_c(S^\iota, q)| > 1$ for some $q \in Q_c(S^\iota)$.

$$
\begin{aligned}
\Delta\mu &\triangleq \mu^\iota - \mu^* \\
&= \mu(L(S^\iota/G)) - \mu(L(S^*/G)) \\
&= [\mathbf{I} - \mathbf{\Pi}(S^\iota)]^{-1}\mathbf{X} - [\mathbf{I} - \mathbf{\Pi}(S^*)]^{-1}\mathbf{X} \\
&= [\mathbf{I} - \mathbf{\Pi}(S^\iota)]^{-1}\left\{[\mathbf{I} - \mathbf{\Pi}(S^*)] - [\mathbf{I} - \mathbf{\Pi}(S^\iota)]\right\} \\
&\quad [\mathbf{I} - \mathbf{\Pi}(S)^*]^{-1}\mathbf{X} \\
&= [\mathbf{I} - \mathbf{\Pi}(S^\iota)]^{-1}\left(\mathbf{\Pi}(S^\iota) - \mathbf{\Pi}(S^*)\right)\mu^*
\end{aligned}
$$

(1) $[\mathbf{I} - \mathbf{\Pi}(S^\iota)]^{-1} \geq 0$. By Taylor series expansion,

$$
[\mathbf{I} - \mathbf{\Pi}(S^\iota)]^{-1} = \sum_{n=0}^{\infty} (\mathbf{\Pi}(S^\iota))^n \tag{6.30}
$$

Since each element of $\mathbf{\Pi}(S^\iota)$ is non-negative, so is each element of $(\mathbf{\Pi}(S^\iota))^n$. Therefore, $[\mathbf{I} - \mathbf{\Pi}(S^\iota)]^{-1} \geq 0$ elementwise.

(2) Let us suppose $\Sigma_c(S^\iota, q_j) = \{\sigma_m, \sigma_n\}$ and $\Sigma_c(S^*, q_j) = \{\sigma_l\}$, for some $j \in \mathcal{I}$, then the j-th element of $(\mathbf{\Pi}(S^\iota) - \mathbf{\Pi}(S^*))\mu^*$ is given by

$$
\begin{aligned}
\Delta_j &= ([0 \cdots \tilde{\pi}_{jm} \cdots \tilde{\pi}_{jn} \cdots 0] - [0 \cdots \tilde{\pi}_{jl} \cdots 0]) \\
&\quad [\cdots \mu_m^* \cdots \mu_l^* \cdots \mu_n^* \cdots]^T \\
&= \tilde{\pi}_{jm}\mu_m + \tilde{\pi}_{jn}\mu_n - \tilde{\pi}_{jl}\mu_l \\
&\leq (\tilde{\pi}_{jm} + \tilde{\pi}_{jn})\mu_\kappa - \tilde{\pi}_{jl}\mu_l \quad (\mu_\kappa = \max\{\mu_m, \mu_n\}) \\
&< (\tilde{\pi}_{jm} + \tilde{\pi}_{jn} - \tilde{\pi}_{jl})\mu_l \quad (\because \ \mu_m < \mu_l) \\
&< 0 \quad (\because \ \tilde{\pi}_{jm} + \tilde{\pi}_{jn} = \tilde{\pi}_{jl})
\end{aligned}
$$

Since $\Delta_j < 0$ for every $j \in \mathcal{I}$, $\Delta\mu < 0$. This contradicts the original hypothesis that the supervisor S^* is optimal, i.e. $\mu(L(S/G)) \leq \mu(L(S^*/G))$ for all supervisors S.

Intuitively, at a given state, a control synthesis algorithm should attempt to enable only the controllable event that leads to the next state with highest performance measure μ, equivalently, disabling the rest of controllable events defined at that state, if any. A recursive synthesis algorithm is first presented and then it is shown that μ is monotonically increasing elementwise on every iteration.

(1) Initialization. Set $\mathbf{\Pi}^0 = \mathbf{\Pi}_p$, then compute $\mu^0 = (\mathbf{I} - \mathbf{\Pi}^0)^{-1}\mathbf{X}$.

(2) Recursion. At k-th iteration, where $k \geq 1$,

   a) μ maximization. For every $q \in Q_c(G)$, identify the event $\sigma^* \in \Sigma_c(G, q)$ such that $q^* = \delta(q, \sigma^*)$ and

$$\mu(L(S^k(\widetilde{\mathbf{\Pi}}^k), q^*)) = \max_{\substack{\sigma \in \Sigma_c(G,q) \\ q' = \delta(q,\sigma)}} \mu(L(S^k(\widetilde{\mathbf{\Pi}}^k), q')) \qquad (6.31)$$

where $S^k(\widetilde{\mathbf{\Pi}}^k)$ is the intermediate supervisor at k-th iteration whose transition is determined by $\widetilde{\mathbf{\Pi}}^k$. Let $\boldsymbol{\sigma}^* = [\sigma_1^* \ \sigma_2^* \ \cdots \ \sigma_n^*]^T$, where the i-th element in $\boldsymbol{\sigma}^*$ is the controllable event left-enabled at state q_i of the plant G according to Equation 6.31.

3) Event Disabling. Disable the event set $\Sigma_c(G,q) - \{\ \boldsymbol{\sigma}^*(q)\}$ for every $q \in Q_c(G)$ and redistribute event cost according to Equation 6.23. This results in a new $\widetilde{\mathbf{\Pi}}^{k+1}$-matrix which consequently produces a new $\mathbf{\Pi}^{k+1}$-matrix.

$$\mathbf{\Pi}^{k+1} = \mathbf{\Pi}^k + \Delta^k \qquad (6.32)$$

where Δ^k records the difference between $\mathbf{\Pi}^{k+1}$ and $\mathbf{\Pi}^{k+1}$, consisting positive and negative event costs corresponding to those kept and disabled controllable events, respectively. The resulting supervisor is $S^{k+1}(\widetilde{\mathbf{\Pi}}^{k+1})$.

4) Measure Computation. Compute $\boldsymbol{\mu}^{k+1} = (\mathbf{I} - \mathbf{\Pi}^{k+1})^{-1}\mathbf{X}$

3. Termination. If $\widetilde{\mathbf{\Pi}}^{k+1} = \widetilde{\mathbf{\Pi}}^k$, then stop.

At each iteration of the recursive algorithm, neither the number of states is increased, nor any additional transition is added in the supervisor $S^k(\widetilde{\mathbf{\Pi}}^k)$ with respect to the plant automaton G. Therefore, $L(S^k(\widetilde{\mathbf{\Pi}}^k)) \subseteq L(G)$.

Theorem 6.4.3 (Monotonicity) *The sequence $\boldsymbol{\mu}^k$, $k = 1, 2, \ldots$, generated recursively by the above algorithm is monotonically increasing.*

Proof. To show that $\{\ \boldsymbol{\mu}^k,\ k \in \mathbb{N}\}$ is a monotonic sequence, it suffices to show that $\Delta\boldsymbol{\mu}^k = \boldsymbol{\mu}^{k+1} - \boldsymbol{\mu}^k \geq 0$ for every $k \in \mathbb{N}$.

$$\begin{aligned}
\Delta\boldsymbol{\mu}^k &= [\mathbf{I} - \mathbf{\Pi}^{k+1}]^{-1}\mathbf{X} - [\mathbf{I} - \mathbf{\Pi}^k]^{-1}\mathbf{X} \\
&= [\mathbf{I} - \mathbf{\Pi}^{k+1}]^{-1} \left[[\mathbf{I} - \mathbf{\Pi}^k] - [\mathbf{I} - \mathbf{\Pi}^{k+1}] \right] \\
&\quad [\mathbf{I} - \mathbf{\Pi}^k]^{-1}\mathbf{X} \\
&= [\mathbf{I} - \mathbf{\Pi}^{k+1}]^{-1}(\mathbf{\Pi}^{k+1} - \mathbf{\Pi}^k)\boldsymbol{\mu}^k \\
&= [\mathbf{I} - \mathbf{\Pi}^{k+1}]^{-1}\Delta^k\boldsymbol{\mu}^k
\end{aligned}$$

(1) $[\mathbf{I} - \mathbf{\Pi}^{k+1}]^{-1} \geq 0$. *By Taylor series expansion,*

$$[\mathbf{I} - \mathbf{\Pi}^{k+1}]^{-1} = \sum_{n=0}^{\infty} \left(\mathbf{\Pi}^{k+1}\right)^n \qquad (6.33)$$

Since each element of $\mathbf{\Pi}^{k+1}$ is non-negative, so is each element of $\left(\mathbf{\Pi}^{k+1}\right)^n$. Therefore, $[\mathbf{I} - \mathbf{\Pi}^{k+1}]^{-1} \geq 0$ elementwise.

(2) $\Delta^k \mu^{k+1} > 0$. For $k \geq 1$, by the recursive algorithm, for any state $q \in Q_c(G)$, there is one and only one left enabled in $S^k(\widetilde{\Pi}^k)$. Let σ_m and σ_n be the only controllable event left enabled on some state $q_l \in Q_c(G)$ at the k-th and $k+1$-th iteration, respectively. Then, only the m-th element and the n-th element of the l-th row are non-zero in Π^k and Π^{k+1}, respectively. So the l-th element of $\Delta^k \mu^{k+1}$ is given by:

$$
\Delta^k \mu^{k+1}(l) = \left(\begin{bmatrix} 0 \\ \vdots \\ \sum_j \tilde{\pi}_{ij}^{k+1} \\ \vdots \\ 0 \\ \vdots \end{bmatrix} - \begin{bmatrix} 0 \\ \vdots \\ 0 \\ \vdots \\ \sum_j \tilde{\pi}_{ij}^{k} \\ \vdots \end{bmatrix} \right)^T
$$

$$
\begin{bmatrix} \mu_1^k & \cdots & \mu_n^k & \cdots & \mu_m^k & \cdots \end{bmatrix}^T
$$

$$
= \sum_j \tilde{\pi}_{ij} (\mu_n^k - \mu_m^k) \text{ (by Equation 6.24)}
$$

$$
> 0
$$

The last inequality holds because σ_m is disabled and σ_n is enabled in the $(k+1)$-th iteration only if μ_n is greater than μ_m, according to the recursive algorithm.

Given that every iteration in the above recursive synthesis algorithm generates an intermediate supervisor S^k in the form of $\widetilde{\Pi}^k$ at the kth iteration and its measure μ^k is monotonically increasing with k, the algorithm converges in a finite number of iterations to yield the μ-optimal supervisor S^*. It has been shown by Fu et al. [2] that complexity of the above optimal algorithm is polynomial in number of the plant automaton states.

6.5 Validation of the Supervisory Control Algorithm

This section validates the supervisory control algorithm on a robotic test bed; the control system architecture is shown in Figure 6.1. The test bed makes use of the Player/Stage[5]. Player is a device server that manages the robot's sensors and actuators, whereas Stage can stimulate a population of mobile robots, sensors and objects in a two-dimensional bitmapped environment. In the simulation experiments, a Pioneer 2 AT robot is equipped with sonar, laser range finder, vision camera, gripper, battery. The DES control module (DCM) communicates with the existing continuous varying control module(CVCM)

---

[5] Player/Stage was developed at University of Southern California and is available under GNU Public License at http://playerstage.sourceforge.net/

of the robot by sending and receiving a set of discrete events, listed in Table 6.1). The DCM is designed to be independent of the underlying physical process such that it provides a mechanism to interact with the CVCM. A DES controller is allowed to plug and play in DCM where the DES controller is loaded in a special format. The CVCM consists of three blocks: continuous-to-discrete (C/D), discrete-to-continuous (D/C), and the CVCM. The CVCM is connected to the Player via a standard TCP socket for sending and receiving formatted messages that encode up-to-date sensor readings and continuously varying commands of reference signals, respectively, at 10 Hz. The control strategy is event-driven, in which the CVCM generates discrete events based on the continuous sensor data received from the Player. The discrete events are transmitted to DCM in a symbolic form. For example, if the robot is in *search* mode and the incoming vision data indicate the presence of a red unit in its view, the CVCM generates a "find red unit" event. In response, DCM reacts immediately by sending out a a discrete event command back to the CVCM according to the currently loaded supervisor; in this case, the commands are either "ignore unit" or "proceed to supply". After receiving a particular event from the DES controller, the CVCM sends out a set of continuous reference signals to Player for completion of this behavior of accordingly maneuvering the robot. The robot continues executing this behavior until the sensor readings trigger the occurrence of a new event.

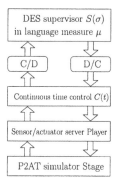

Fig. 6.1. Pioneer 2 AT DES simulation block diagram

The experimental scenario consists of a single robot performing logistic supply and combat operation in a simulated battle field. There are two friendly units (represented by red and green colored circles) and one enemy (represented by blue circle), which are at stationary locations in the field. The robot does not have prior knowledge of the environment. When a "start mission" signal is received, the robot randomly searches for a red or green unit. When finding a unit, the robot can either proceed to supply or ignore the unit and keep on searching. It may also encounter an enemy during the course of searching. The robot may decide either to avoid or to fight the enemy. In both cases, there are chances that the robot may fail to supply or lose the fight.

A gripper failure is modelled to signify the robot's failure to complete the task of supplying units. However, fighting with enemy is still possible. In addition, during the mission, the battery consumption rate changes according to the robot's current actions. For example, the robot's power consumption is higher during the fight with the enemy than supplying the units. The robot needs to return to its base (represented by a large pink circle) to be recharged before the battery voltage drops below a certain level, or the robot is considered to be lost. After each successful return to the base, the robot is reset to normal conditions including full battery charge and normal gripper status. If the robot is incapacitated either due to battery over-drainage or damage in a battle, the "death toll" is increased by one. In both cases, the mission is automatically restarted with a new robot.

Due to the independence of the robot operation, battery consumption, and occurrence of gripper failures, the entire interaction between robot and environment is modelled by three different submodels, as shown in Figure 6.2 (a) and (b). The blue-colored and red-colored states are *good* marked states in Q_m^+ and *bad* marked states in Q_m^-, respectively. Then the models are composed by synchronous composition operator defined below to generate the integrated plant model $G = G_1 \| G_2 \| G_3$.

Definition 6.5.1 *Given two finite state automata, $G_1 = (Q_1, \Sigma_1, \delta_1, q_{0,1}, Q_{m,1})$ and $G_2 = (Q_2, \Sigma_2, \delta_2, q_{0,2}, Q_{m,2})$, synchronous composition of G_1 and G_2, denoted $G_1 \| G_2 = (Q, \Sigma, \delta, q_0, Q_m)$, is defined as an automaton with $Q = Q_1 \times Q_2, \Sigma = \Sigma_1 \cup \Sigma_2, q_0 = (q_{0,1}, q_{0,2}), Q_m = Q_{m,1} \times Q_{m,2}$ and for every $q = (q_1, q_2) \in Q, \sigma \in \Sigma$.*

$$\delta(q, \sigma) = \begin{cases} (\delta_1(q_1, \sigma), \delta_2(q_2, \sigma)) & \sigma \in \Sigma_1 \cap \Sigma_2 \\ (\delta_1(q_1, \sigma), q_2) & \sigma \in \Sigma_1 - \Sigma_2 \\ (q_1, \delta_2(q_2, \sigma)) & \sigma \in \Sigma_2 - \Sigma_1 \\ undefined & otherwise \end{cases} \tag{6.34}$$

After eliminating the inaccessible states, the discrete-event model of the plant automaton G consists of 139 states; and there are 21 events, as listed in Table 6.1. The event cost matrix $\tilde{\Pi}$ is then identified by Monte Carlo simulation over 1200 missions according to the parameter identification procedure; convergence of selected non-zero elements in $\tilde{\Pi}$-matrix is demonstrated in Figure 6.3. For those states that have more than one controllable event defined, the probabilities of occurrence are assumed to be equally distributed. A vast majority of the plant states are unmarked; consequently, the corresponding elements, χ_i, of the characteristic vector are zero. Negative characteristic values, χ_i are assigned to the bad marked states. For example, the states in which the robot is dead due to either losing the battle to the enemy or running out of battery is assigned the negative value of -1. Similarly, positive

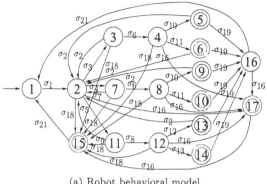

(a) Robot behavioral model

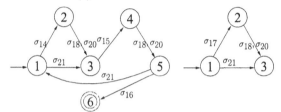

(b) Battery failure model and gripper failure model

Fig. 6.2. Finite State Models of Robot Operation

Table 6.1. List of discrete events

Σ	Description	Σ	Description
σ_1	start mission	σ_{12}	win the fight
σ_2	search	σ_{13}	loose the fight
σ_3	find blue unit	σ_{14}	battery power medium
σ_4	find pink unit	σ_{15}	battery power low
σ_5	find enemy	σ_{16}	battery power dead
σ_6	proceed to supply	σ_{17}	detected gripper fault
σ_7	ignore unit	σ_{18}	abort mission
σ_8	fight enemy	σ_{19}	return
σ_9	avoid enemy	σ_{20}	ignore anomaly
σ_{10}	finish supply	σ_{21}	return successfully
σ_{11}	fail supply		

values are assigned to good states. For example, the state in which the robot wins a battle is assigned 0.5 and the state of successfully providing supplies is assigned 0.3. Using the recursive synthesis algorithm in Section 6.4, the μ-optimal supervisor S^* is then synthesized. The optimal DES control algorithm converges at the fourth iteration, as listed in Table 6.2. For the purpose of performance comparison, two additional supervisors, S_1 and S_2 under the following specifications are designed.

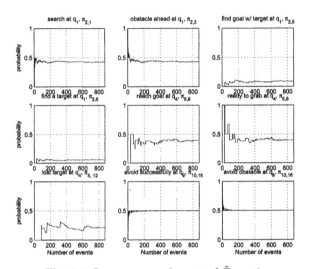

Fig. 6.3. Some non-zero elements of $\tilde{\Pi}$-matrix

Table 6.2. Iteration of μ synthesis

k-th Iteration	μ^k
0	-1.7834
1	2.8306
2	4.5655
3	4.5696
4	4.5696

The specifications of Controller S_1 are as follows:

1. Avoid enemy when the battery power is not below medium;
2. Abort all operation when the battery power is low;

3. If there is a gripper failure, do not supply a discovered unit or abort supply if the supply is on-going.

The specifications of Controller S_2 are as follows:

1. Abort all operation if battery power is not below medium;
2. Abort all operation if a gripper failure is detected.

The conventional DES controllers, S_1 and S_2, have 109 and 54 states, respectively, and 400 missions were simulated for the open loop plant and each of the three DES controllers: μ-optimal supervisor S^*, and the conventional S_1 and S_2. The statistics of the simulation results are summarized in Table 6.3, where the supervised robot yields higher performance under S^* than under S_1 and S_2 or the null supervisor (i.e., the unsupervised plant G). In the last row of Table 6.3, the cumulative performance of robot operations, with the initial state q_1, is obtained by summing the language measure over 400 missions as: $\sum_{i=1}^{400} \mu_1(i)$.

During these 400 missions, S^* decides to *proceed to supply* for 482 times, three times more than S_1 or S_2. In addition, the probability of deciding to *proceed to supply*, when the robot sees a unit, is much higher than S_1 or S_2. However, the price of this decision is that the robot is likely to drain out the battery energy and therefore may risk being immobile in the middle of the mission. The μ-optimal supervisor S^* also decides to fight many more times (167) than any other supervisor since the reward to win a battle is large ($\chi = 0.5$). Certainly, the number of robots lost under S^* is also higher (48) than that under other supervisors because S^* has to expose the robot more often to the enemy attack to win the battles. On the average, S^* outperforms G, S_1 and S_2 in measure $\mu = (\mathbf{I} - \mathbf{\Pi})^{-1} \mathbf{X}$. The cumulative performance of S^* is found to be superior to the cumulative performance of two supervisors and the null supervisor (i.e., open loop or unsupervised robot) over 400 missions as seen in Figure 6.4. The oscillations in the performance profile, as a function of the number of missions, are attributed to unavailability of the robot resulting from drained battery or damage in the battle.

Table 6.3. Simulation statistics of 400 missions

Items	G	S_1	S_2	S^*
proceed to supply	150	134	137	482
# of units found	434	408	367	556
	34.56%	32.84%	37.33%	**86.69%**
finish supply	112	88	97	362
win enemy	37	33	22	69
fight enemy	65	70	68	167
unit lost	26	19	14	48
μ	-1.7834	-1.4428	-1.8365	**4.5696**
$\sum_{i=1}^{400} \mu_1(i)$	-20.25	-19.45	-27.3	**45.25**

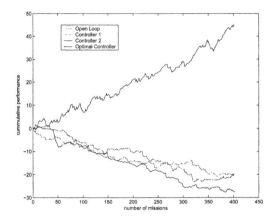

Fig. 6.4. Cumulative performance comparison of all controllers

6.5.1 Optimal control synthesis under a given scenario

A scenario of object collection by robot(s) is presented in Chapter 5, where an ad hoc supervisor has been carefully designed based on experience and engineering judgement. This chapter synthesizes an optimal supervisor based on the language measure μ and compares its performance with that of the ad hoc supervisor designed in Chapter 5. The optimal supervisor algorithm converges after 2 iterations as seen in Table 6.4 and Table 6.5 under simulation and robotic test bed experiments, respectively. Both simulation and experimental results exhibit (elementwise) monotonicity at each iteration, where μ_3 and μ_{13} remain unchanged at each iteration. By comparing the corresponding elements of $\widetilde{\Pi}$-matrix, it is observed that the probabilities of visiting these two states are zero, i.e., they were never visited. Consequently, the 3rd row and the 13th row of $(\mathbf{I} - \mathbf{\Pi}^k)^{-1}$ contain one and only one none zero element at the 3rd and the 13th elements, respectively. In other words, the discrete-event plant model could have been further refined by deleting state 3 and state 13 from the unsupervised plant model G. It is interesting to note that performance of the ad hoc supervisor is very close to that of the optimal supervisor. However, in more complex systems, it is expected that an ad hoc supervisor, however carefully designed, would not be able to compete with the optimal supervisor.

6.5.2 Synopsis of the DES control synthesis procedure

This section provides a synopsis of the procedure for the synthesis of discrete-event supervisory (DES) control systems in terms of the language measure.

1. Formulation of an unsupervised plant automaton model of the process dynamics;

Table 6.4. Iteration of μ synthesis with $(\widetilde{\mathbf{\Pi}}, \mathbf{X})$ obtained in simulation

μ Elements	μ^0	μ^1	μ^2
μ_0	**0.0799**	**0.1349**	**0.1349**
μ_1	0.0840	0.1420	0.1420
μ_2	0.1578	0.2781	0.2781
μ_3	0	0	0
μ_4	0.3667	0.4071	0.4071
μ_5	0.7218	0.7670	0.7670
μ_6	0.3937	0.4389	0.4389
μ_7	0.8798	0.9349	0.9349
μ_8	0.4798	0.5349	0.5349
μ_9	0.0694	0.1173	0.1173
μ_{10}	-0.1202	-0.0651	-0.0651
μ_{11}	0.7598	0.8074	0.8074
μ_{12}	0.4144	0.4620	0.4620
μ_{13}	-1.0000	-1.0000	-1.0000
μ_{14}	0.3514	0.4735	0.4735
μ_{17}	0.0664	0.1843	0.1843
μ_{16}	-0.2002	-0.1501	-0.1501

Table 6.5. Iteration of μ synthesis with $(\widetilde{\mathbf{\Pi}}, \mathbf{X})$ obtained in experiment

μ Elements	μ^0	μ^1	μ^2
μ_0	0.1640	0.1987	0.1987
μ_1	0.1726	0.2091	0.2091
μ_2	0.2737	0.3382	0.3382
μ_3	0	0	0
μ_4	0.5114	0.5418	0.5418
μ_5	0.8700	0.9013	0.9013
μ_6	0.5090	0.5403	0.5403
μ_7	0.9640	0.9987	0.9987
μ_8	0.5640	0.5987	0.5987
μ_9	0.1562	0.1892	0.1892
μ_{10}	-0.0360	-0.0013	-0.0013
μ_{11}	0.9158	0.9487	0.9487
μ_{12}	0.5358	0.5687	0.5687
μ_{13}	-1.0000	-1.0000	-1.0000
μ_{14}	0.4798	0.5436	0.5436
μ_{15}	0.2229	0.2818	0.2818
μ_{16}	-0.1000	-0.0681	-0.0681

2. Construction of a high fidelity simulator of the physical process;
3. Extensive simulation to obtain the $\widetilde{\Pi}$-matrix;
4. Assignment of the $\mathbf{X}$-vector based on the user's perception of the desired plant performance;
5. Design of a supervisor having a high μ value of the supervised plant automaton;
6. Deployment of the supervisor in the simulator or the physical process and updating the $\widetilde{\Pi}$-matrix through the real-environment experiments;
7. Re-design the supervisory algorithm according to the optimal μ synthesis procedure if the elements of the $\widetilde{\Pi}$-matrix deviate from the original values above a specified threshold.

6.6 Related work on Q-learning

This section presents the concept of Q-learning [11] that is widely used for reinforcement learning in behavior-based robotics [1] for its algorithmic simplicity and ease of transformation from a state function to an optimal control policy. In addition, similar to the approach presented in this chapter, it does not require a world model. It is well suited for use in stationary, single-agent, fully observable environments that are modeled by Markov decision processes (MDPs). However, it often performs well in environments that violate these assumptions. Mahadevan and Connell [3] have used Q-learning to teach a behavior-based robot how to push boxes around a room without getting stuck. The mission is heuristically divided into three behaviors: *finding a box, pushing a box*, and *recovering from stalled situations*. The results show that the mission decomposition method is capable of learning faster than a method that treats the mission as a single behavior. Martinson *et al* [4] demonstrated by both simulation and robotic experiments that Q-learning can be used for robot behavioral selection to optimize the overall mission in terms of Q-value. The basic principles of Q-learning are briefly described below.

A *Markov decision process* is a tuple $M = (S, A, P, R)$, where S is the set of states; A is the set of actions; $P : S \times A \times S \rightarrow [0, 1]$ the transition probability function; and $R : S \times A \rightarrow \mathbb{R}$ is an expected reward function. Let $\pi : S \rightarrow A$ be a mapping between what has happened in the past and what has to be done at the current state. The value of the policy π, starting at state s, is evaluated as:

$$V^\pi(s) \equiv E\left(\sum_{t=0}^{\infty} \gamma^t r_t\right) \tag{6.35}$$

$$= R(s, a) + \gamma \sum_{s' \in S} P(s, a, s') V^\pi(s') \tag{6.36}$$

where r_t is the reward for taking a transition at time t; γ^t is the discount factor at time t for $\gamma^t \in [0, 1)$; $R(s, a)$ is the expected instantaneous reward by action a at state s; and $P(s, a, s')$ is the probability of making a transition from state s to state s' by action a.

The objective is to find an optimal control policy π^* that maximizes the future reward with a discount factor γ. One of the important results, used in the Q-learning, is that there exists an optimal stationary policy π^* that maximizes $V^\pi(s)$ for all states s, where the optimal policy V^{π^*} is denoted by V^*.

Let $Q(s, a)$ denote the discounted reinforcement of taking action a at state s following a policy π. Using the notation of [11], $Q^*(s, a)$ is the value of $Q(s, a)$, starting at state s and taking action a and from there on following the optimal policy π^*. Therefore,

$$Q^*(s, a) = R(s, a) + \gamma \sum_{s' \in S} P(s, a, s') V^*(s) \tag{6.37}$$

$$\pi^* = \arg \max_a Q^*(s, a) \tag{6.38}$$

At each state, the best policy π^* is to take the action with the largest Q-value, i.e., $V^*(s) = \max_a Q^*(s, a)$. The basic idea of Q-learning is to maintain an estimate $\hat{Q}(s, a)$ of $Q^*(s, a)$. The on-line Q-learning update rule is:

$$\hat{Q}(s, a) = \hat{Q}(s, a) + \alpha(r + \gamma \max_{a'} \hat{Q}(s', a') - \hat{Q}(s, a)) \tag{6.39}$$

where α is the learning rate. In [11] [12], it has been shown that the Q values will converge with probability 1 to Q^* if each action is executed in each state an infinite number of times on an infinite run and α is decayed appropriately.

6.7 Conclusions

This chapter presents optimal supervisory control of robot behavior based on the discrete-event language measure μ [10] [8] as described in detail in Chapter 1. A recursive supervisor synthesis algorithm is provided as an extension of the earlier work of Fu et al. [2]. A real-world robot simulation scenario has been developed for experimental validation of this new concept of discrete event supervisory (DES) control. The results of simulation experiments validate that a DES controller, designed by the recursive synthesis algorithm, exceeds the μ-measure and also has significantly better performance than two other supervisors, designed in conventional way, and the null supervisor (i.e., the unsupervised plant).

The chapter also compares the proposed language measure (μ)-based approach of DES control with the Q-learning method, where the reward function $R(s, a)$ is defined for every transition at each state of the Markov decision process based on the designer's perception, similar to the characteristic value (i.e., the χ value) assigned to every state of the plant automaton G in μ_i

in the construction of the language μ. Conceptually, a weight is assigned to each transition in Q-learning whereas a weight is assigned to each state in the μ-selection process. However, while complexity of the μ-based DES control is polynomial in number of plant automaton states, the complexity of the Q-learning method increases exponentially with the number of states n and the number of actions (events) m. A summary of comparison between Q-learning and μ-selection is given in Table 6.6.

Table 6.6. Comparison of Q-learning and μ-selection

Items	Q-learning	μ-selection
Modeling	$M = (S, A, T, R)$	$G = (Q, \Sigma, \delta, q_0, Q_m)$
Control policy	MDP	DES
Objective	$\arg\max_a Q^*(s, a)$	$\max_{s \in L(G)} \mu(s)$
Transition probability	required $T(s, a, s')$	required $\tilde{\pi}(q, \sigma)$
Reward	on transition $r(s, a)$	on state $\chi(q)$
Discount factor γ	yes(ad hoc)	no
Learning rate α	yes(ad hoc)	no
Online adaptation to dynamic environment	recursive	(ϵ, δ)-threshold, redesign
Model complexity	exponential in S and A	polynomial in Q

References

1. R. C. Arkin, *Behavior-based robotics*, MIT Press, 1998.
2. J. Fu, A. Ray, and C.M. Lagoa, *Unconstrained optimal control of regular languages*, Automatica **40** (2004), no. 4, 639–648.
3. S. Mahadevan and J. Connell, *Automatic programming of behavior-based robots using reinforcement learning*, Proceedings of AAAI-91, 1991, pp. 768–773.
4. E. Martinson, A. Stoytchev, and R. C. Arkin, *Robot behavioral selection using q-learning*, IEEE International Conference on Robots and Systems (Lausanne), September 2002.
5. M. Pradhan and P. Dagum, *Optimal monte carlo estimation of belief network inference*, Twelfth Conference on Uncertainty in Artificial Intelligence (Portland, OR), 1996, pp. 446–453.
6. P.J. Ramadge and W.M. Wonham, *Supervisory control of a class of discrete event processes*, SIAM J. Control and Optimization **25** (1987), no. 1, 206–230.

7. A. Ray and S. Phoha, *Signed real measure of regular languages for discrete-event automata*, Int. J. Control **76** (2003), no. 18, 1800–1808.
8. A. Surana and A. Ray, *Signed real measure of regular languages*, Demonstratio Mathematica **37** (2004), no. 2, 485–503.
9. X. Wang, *Quantitative measure of regular languages for supervisory control of engineering applications*, Ph.D. thesis, The Pennsylvania State University, December 2003.
10. X. Wang and A. Ray, *A language measure for performance evaluation of discrete-event supervisory control systems*, Applied Mathematical Modelling **28** (2004), no. 9, 817–833.
11. C.J.C.H. Watkins, *Learning from delayed rewards*, Ph.D. thesis, King's College, Cambridge, UK, 1989.
12. C.J.C.H. Watkins and P. Dayan, *Q-learning*, Machine Learning **8** (1992), no. 3, 279–292.

7

Optimal Discrete Event Control
of Gas Turbine Engines

Murat Yasar[1], Jinbo Fu[2], and Asok Ray[3]

[1] The Pennsylvania State University myasar@psu.edu
[2] The Pennsylvania State University jinbofu@adelphia.net
[3] The Pennsylvania State University axr2@psu.edu

Summary. This chapter presents an application of the recently developed theory of optimal Discrete Event Supervisory (DES) control that is based on a signed real measure of regular languages described in Chapter 1. The DES control techniques are validated on an aircraft gas turbine engine simulation test bed. The test bed is implemented on a networked computer system in which two computers operate in the client-server mode. Several DES controllers have been tested for engine performance and reliability. Extensive simulation studies on the test bed show that the optimally designed supervisor yields the best performance.

Key words: Discrete Event Supervisory Control, Gas Turbine Engine Control, Optimal Control, Language Measure

7.1 Introduction

The concept of Discrete Event Supervisory (DES) control is potentially very useful for health management and intelligent control at an upper level of hierarchy, especially for large-scale interconnected systems; DES control also plays a critical role in hybrid control systems where the information generated from continuous-time sensors causes the events of Deterministic Finite State Automata (DFSA) model to occur. For hierarchical control, continuously-varying and/or discrete-event signals at the lower level trigger the discrete events belonging to the alphabet of the higher-level DES control [4].

Discrete-event dynamical behavior of physical plants is often modeled as regular languages that can be realized by finite-state automata [4]. Since the sublanguages of a controlled physical plant may be different under different supervisors, the partially ordered set of sublanguages requires a quantitative measure for total ordering of their respective performance. This issue has been addressed in Chapter 1 that has formulated a signed measure of regular languages based on the work reported by Wang and Ray [7], Ray and Phoha [5], and Surana and Ray [6]. A theory of optimal control of regular languages has been developed in Chapter 2 based on the work recently reported in a series of papers [3] [2] [1].

This chapter presents an application of the optimal Discrete Event Supervisory (DES) control theory, which addresses state-based optimal supervisory control of a commercial-scale gas turbine engine without event disabling cost. The performance index of the optimal control policy is a signed language measure of the supervised sublanguage, which is expressed in terms of a state transition cost matrix and a characteristic vector that are parameters for the language measure. Apparently, this is the first time that the theory of optimal DES control has been applied to a complex dynamical system such as a gas turbine engine.

A simulation test bed has been constructed for studying the low-frequency transient behavior of typical turbofan engines under different supervisory control policies. A new Deterministic Finite State Automaton (DFSA) plant model is developed to capture the discrete-event dynamic behavior of a generic gas turbine engine. A lumped parameter model at the component level represents the continuously varying dynamics of engine cycle operations. The test bed executes one pass within the digital controller's sampling time and thermodynamic states are assumed to be in equilibrium after each pass through the simulation of a low-frequency transient turbofan engine. In the model of engine simulation, volume dynamics and airflow storage effects, which are high frequency phenomena, are not included. The continuously varying part of the engine model is driven by six actuators; there are ten parameters that describe the health of the engine. For this application, the efficiency of the high pressure turbine is chosen as the indicator of engine health. Since the continuously varying part of the engine model does not include various aspects of engines abnormal operations, such as low oil pressure, high bearing vibration, and foreign object impact, which are of paramount interest to the pilot. For proper decision and control, this information is necessary, and was obtained by running various engine scenario simulations and compiling information regarding pilot experience. The resulting DFSA plant model serves to synthesize an optimal discrete event supervisory (DES) control system for engine operations.

The engine simulation test bed is implemented on a networked system, where two computers operate in the client-server mode. The plant (i.e., engine operation) model is hosted in the client computer and executes the engine simulation. The control computer is the server and executes the tasks of the DES control and other ancillary functions such as information display. Several DES controllers, including the unsupervised plant (i.e., the engine without DES control), have been tested for comparison of engine performance and reliability.

The chapter is organized in six sections including the present one. Section 7.2 reviews the salient concepts of the language measure. Section 7.3 summarizes the DES control techniques. Section 7.4 discusses the implementation of DES control on the engine simulation. In Section 7.5, the experiments carried out are discussed, and the experimental results are examined. The chapter is summarized and concluded in Section 7.6.

7.2 Brief Review of the Language Measure

This section reviews the previous work on language measure [7] [6]. It provides the background information necessary to develop a performance index and an optimal control policy.

Let the dynamical behavior of a physical plant be modeled as a deterministic finite state automaton (DFSA) $G_i \equiv (Q, \Sigma, \delta, q_i, Q_m)$ where Q is the finite set of states with $|Q| = n$ where $q_i \in Q$ is the initial state. Σ is the (finite) alphabet of events with $|\Sigma| = \ell$; Σ^* is the set of all finite-length strings of events including the empty string ϵ. The (possibly partial) function of $\delta : Q \times \Sigma \to Q$ represents state transitions and $\hat{\delta} : Q \times \Sigma^* \to Q$ is an extension of δ; and $Q_m \subseteq Q$ is the set of *marked* (also known as *accepted*) states.

Definition 7.2.1 *A DFSA G_i, initialized at $q_i \in Q$, generates the language $L(G_i) \equiv \{s \in \Sigma^* : \hat{\delta}(q_i, s) \in Q\}$ and its marked sublanguage $L_m(G_i) \equiv \{s \in \Sigma^* : \hat{\delta}(q_i, s) \in Q_m\}$.*

Following Wang and Ray [7] and Surana and Ray [6], it is possible to construct a signed real measure $\mu : 2^{\Sigma^*} \to \Re \equiv (-\infty, \infty)$ that allows quantitative evaluation of every event string $s \in \Sigma^*$ based on state-based decomposition of $L(G_i)$ into null (i.e., $L^0(G_i)$), positive (i.e., $L_m^+(G_i)$ and negative $L_m^-(G_i)$) sublanguages of $L(G_i)$.

Definition 7.2.2 *The language of all strings starting at $q_i \in Q$, and terminating at $q_j \in Q$ is denoted as $L(q_i, q_j)$.*

That is, $L(q_i, q_j) \equiv \{s \in L(G_i) : \hat{\delta}(q_i, s) = q_j\}$.

Definition 7.2.3 *The characteristic function that assigns a signed real weight to each state q_j is defined as:* $\chi : Q \to [-1, 1]$ *such that*

$$\chi_j \equiv \chi(q_j) \in \begin{cases} [-1, 0) & \text{if } q_j \in Q_m^- \\ \{0\} & \text{if } q_j \notin Q_m \\ (0, 1] & \text{if } q_j \in Q_m^+ \end{cases}$$

The $(n \times 1)$ characteristic vector is denoted as:

$$\tilde{\chi} \equiv [\chi_1 \ \chi_2 \ \cdots \ \chi_n]$$

Definition 7.2.4 *The event cost is conditioned on a DFSA state at which the event is generated, and is defined as $\tilde{\pi} : \Sigma^* \times Q \to [0, 1]$ such that $\forall q_j \in Q$, $\forall \sigma_k \in \Sigma$, $\forall s \in \Sigma^*$,*

- $\tilde{\pi}[\sigma_k | q_j] = 0$ *if* $\delta(q_j, \sigma_k)$ *is undefined;* $\tilde{\pi}[\epsilon | q_j] = 1$;
- $\tilde{\pi}[\sigma_k | q_j] \equiv \tilde{\pi}_{jk} \in [0, 1)$; $\Sigma_k \tilde{\pi}_{jk} < 1$;
- $\tilde{\pi}[\sigma_k s | q_j] = \tilde{\pi}[\sigma_k | q_j] \tilde{\pi}[s | \delta(q_j, \sigma_k)]$.

The $(n \times m)$ event cost matrix is denoted as: $\tilde{\Pi} = [\tilde{\pi}_{ij}]$.

The cost of an event σ_j including a state transition from q_i to q_k is denoted as $\tilde{\pi}_{ij}^k$

Definition 7.2.5 *The state transition cost of DFSA is defined as a function $\pi : Q \times Q \to [0, 1)$ such that $\forall q_j, q_k \in Q$,*

$$\pi(q_k | q_j) = \sum_{\sigma \in \Sigma : \delta(q_j, \sigma) = q_k} \tilde{\pi}(\sigma | q_j) \equiv \pi_{jk}$$

and $\pi_{jk} = 0$ if $[\sigma \in \Sigma : \delta(q_j, \sigma)] = \emptyset$. The $n \times n$ state transition cost matrix, denoted as Π, is defined as:

$$\Pi = \begin{bmatrix} \pi_{11} & \pi_{12} & \cdots & \pi_{1n} \\ \pi_{21} & \pi_{22} & \cdots & \pi_{2n} \\ \vdots & \vdots & \ddots & \vdots \\ \pi_{n1} & \pi_{n2} & \cdots & \pi_{nn} \end{bmatrix}$$

Now it is possible to define the measure of any sublanguage of the plant language $L(G_i)$ in terms of the signed characteristic function χ and the non-negative event cost $\tilde{\pi}$.

Definition 7.2.6 *The signed real measure μ of a singleton string set $\{s\}$ is defined as: $\mu(\{s\}) \equiv \chi(q_j)\tilde{\pi}(s|q_i) \ \forall s \in L(q_i, q_j) \subseteq L(G_i)$.*
The signed real measure of $L(q_i, q_j)$ is defined as:

$$\mu(L(q_i, q_j)) \equiv \left(\sum_{s \in L(q_i, q_j)} \mu(\{s\}) \right)$$

The signed real measure of a DFSA G_i, initialized at the state $q_i \in Q$, is defined as

$$\mu_i \equiv \mu(L(G_i)) = \sum_j \mu(L(q_i, q_j)).$$

The $n \times 1$ real signed measure vector is denoted as:

$$\bar{\mu} \equiv [\mu_1 \ \mu_2 \ \cdots \ \mu_n]^T$$

Wang and Ray [7] have shown that the measure of the language $L(G_i)$, where $G_i = (Q, \Sigma, \delta, q_i, Q_m)$ can be expressed as: $\mu_i = \sum_j \pi_{ij}\mu_j + \chi_i$. Equivalently, in vector notation: $\bar{\mu} = \Pi\bar{\mu} + \bar{\chi}$. Since Π is a contraction operator, the measure vector $\bar{\mu}$ is uniquely determined as:

$$\bar{\mu} = [I - \Pi]^{-1}\bar{\chi}.$$

7.3 Optimal Discrete Event Supervisory Control

Fu et al. [3] have introduced the concept of unconstrained optimal control of regular languages based on a given language measure. Starting with the (regular) language of an open-loop plant, the optimal control algorithm maximizes the performance of a sublanguage without any further constraints. This optimal control technique is designed for plants with insignificant event disabling cost. The state-based optimal control policy is obtained by selectively disabling controllable events to maximize the measure of the controlled plant language without any constraints. The first iteration of the optimal control algorithm disables selected controllable events and subsequent iterations may re-enable some of the previously disabled controllable events. Synthesis of the optimal control policy requires at most n iterations, where n is the number of states of the DFSA model. Each iteration solves a set of n simultaneous linear algebraic equations. It is also shown that computational complexity of the control synthesis is polynomial in the number of plant states [3]. The construction of the optimal control law that maximizes the performance of the controlled language of the DFSA for any initial state $q \in Q$ is described below.

Let G be the DFSA plant model without any constraint of operational specifications and the state transition cost matrix of the open loop plant be: $\Pi^{plant} \in \Re^{n \times n}$, and the characteristic vector be: $\bar{\chi} \in \Re^n$. Then, the performance vector is given as: $\mu = (I - \Pi)^{-1}\bar{\chi}$, where the jth element μ_j of the vector μ is the performance of the language if the starting state is q_j. As the state q_j is reached, the plant is expected to yield bad performance in the future if $\mu_j < 0$. Intuitively, the control system should attempt to prevent the automaton from reaching q_j by disabling all controllable events that lead to this state. Therefore, the optimal control algorithm starts with disabling all controllable events that lead to every state q_j for which $\mu_j < 0$.

Let us denote the Π-matrix and the performance vector at the kth iteration with the superscript k and start with iteration $k = 0$. Disabling of all controllable events leading to the state q_j is equivalent to reducing all elements of the corresponding j^{th} column of the Π^0-matrix. In the next iteration, i.e., $k = 1$, the updated cost matrix Π^1 is obtained as: $\Pi^1 = \Pi^0 - \Delta^0$ where $\Delta^0 \geq 0$, the inequality being implied element wise, is composed of event costs corresponding to all controllable events that have been disabled. Then, $\mu^0 \leq \mu^1 \equiv [I - \Pi^1]^{-1}\bar{\chi}$. Although all controllable events leading to every state corresponding to a negative element of μ^1 are disabled, some of the controllable events that were disabled at $k = 0$ may now lead to states corresponding to positive elements of μ^1. Performance is enhanced further by re-enabling these controllable events. For $k \geq 1$, $\Pi^{k+1} = \Pi^k + \Delta^k$ where $\Delta^k \geq 0$ is composed of all re-enabled controllable events at k.

Starting with $k = 0$ and $\Pi^0 \equiv \Pi^{plant}$, the control policy is constructed by the following two-step procedure [3]:

Step 1: For every state q_j for which $\mu_j^0 < 0$, disable controllable events leading to q_j. Now, $\Pi^1 = \Pi^0 - \Delta^0$, where $\Delta^0 \geq 0$ is composed of event costs corresponding to all controllable events that have been disabled at $k = 0$.

Step 2: Starting with $k = 1$, if $\mu_j^k \geq 0$, re-enable all controllable events leading to q_j, which were disabled in Step 1. The cost matrix is updated as: $\Pi^{k+1} = \Pi^k + \Delta^k$ for $k \geq 1$, where $\Delta^k \geq 0$ is composed of event costs corresponding to all currently re-enabled controllable events. The iteration is terminated if no controllable event leading to q_j remains disabled for which $\mu_j^k \geq 0$. At this stage, the optimal performance is $\mu^* \equiv [I - \Pi^*]^{-1}\bar{\chi}$.

7.4 Implementation of the Optimal DES Control Concept

This section presents an application of the optimal discrete event supervisory (DES) control for real-time operation of a gas turbine engine. The plant under DES control in the simulation test bed is a nonlinear dynamic model of the engine together with its continuously varying multivariable controller. With the proper inputs of the Power Lever Angle (PLA) and ambient conditions, a FORTRAN program simulates both steady-state and transient operations of the gas turbine engine in the continuous setting. The objectives are to demonstrate efficacy of DES control for: (1) Structural damage reduction and life extension of aircraft engines with DES control interference; and (2) Decision making and mission planning optimization.

7.4.1 Architecture of the DES Engine Controller

The DES control is implemented in the C++ environment around the existing engine simulation code that is written in FORTRAN. The plant model code is a stand-alone program with its own continuous-time gain-scheduled robust controller that is kept unaltered. The C++ wrapper of the simulation code takes over the major inputs and outputs of interest and makes it transparent to the end user as if the entire simulation runs in the C++ environment. The advantage of working in the C++ environment is the convenient utilization of the standard message Application Programming Interface (API) communication routines. In addition, all other functions (e.g., Event generator, Action Generator, and Supervisor), are implemented in C++.

Figure 7.1 shows the architecture of DES control as implemented in the simulation test bed. The DES control is implemented on a pair of networked computers that operate in the server-client mode. The engine simulation program of the aircraft engine and the Action Generator reside in the client computer. The Event Generator and Supervisor are located in the server computer. The pilot commands are entered on the server computer. The server and client communicate with each other over the communication network via API messages as seen in Figure 7.1. Note that the information Fusion module at the bottom of Figure 7.1 requires ancillary (e.g., oil pressure and bearing vibration) information in addition to the conventional plant sensor data (e.g., gas temperature and engine shaft speed).

The Action Generator converts the discrete-event symbols of supervisor commands into continuous signals that are inputs to the plant model. The control commands (e.g., flight parameter modifications, compensating throttle inputs, mission abortion, and Power Lever Angle input) are passed through the Message API communication routine to the Action Generator on the client side. The Action Generator converts control commands from the supervisor into necessary simulation input. Similarly, the Event Generator converts the plant sensor signals and other pertinent information (e.g., engine operational

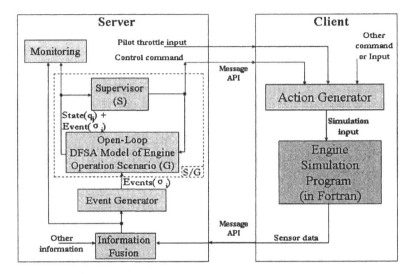

Fig. 7.1. Architecture of the DES control system

data) into event symbols that are inputs to the supervisor. In essence, the role of Event Generator is fusion of heterogeneous information and real-time expression of the relevant part in the language of the supervisor. The sensor signals are processed with built-in information (e.g. threshold values and fault detection logic) to generate discrete events as the inputs to the supervisor. A copy of the DFSA plant model, located in the supervisor, serves as the state estimator.

7.4.2 DFSA Model of Gas Turbine Engine

The open-loop discrete event dynamics are modeled as a DFSA based on the postulated engine operation scenario. The model may vary for different mission scenarios. The DFSA plant model assumes that a military aircraft equipped with a single turbofan jet engine is carrying out a routine surveillance mission. Abortion of the mission is allowed at certain states when an anomaly is detected in the engine. Three major anomalies are considered: Low Oil Pressure, High Bearing Vibration and High Fatigue crack Damage Increment.

The plant model has 42 states, of which four are marked states and the alphabet consists of 16 events, of which four are controllable events [4]. Table 7.1 lists the marked states and controllable events.

Table 7.1. Marked States and Controllable Events

Marked States	Controllable Events
42 - Mission aborted on ground	**a** - Start engine
14 - Mission aborted off ground	**f** - Request for abortion
16 - Unexpected engine halt	**i** - Request flight parameter change
15 - Mission successful	**p** - Maintain current condition

7.4.3 Implementation Details

The 'client-side' in this application consists of a FORTRAN based engine simulation, and a C++ based wrapper program. The wrapper program communicates information between the simulation and server-side program in C++, located on another PC. An 'Action Generator' has been integrated in the simulation to interpret the commands of the DES controller. In addition, the simulation can also respond to inputs from the keyboard during run-time intentionally generating desired events to test the DFSA model on the server side. The Graphic User Interface (GUI) on the server-side controls the communication set-up and starts the client-side simulation.

The variables in the message structure represent simulation parameters that are changing according to the dynamics of the simulation, and are sent to the server after every fixed number of simulation sampling intervals. The software uses 'Message API' interface to establish connection structure and function calls. This interface enables communication of data over a network between client and server.

The role of an Action Generator is to interpret the control action of the discrete event controller and to generate proper inputs for the engine simulation. One of the control actions is to change simulation parameters in order to improve system (simulation) performance. The choice of algorithm is subjective, and the one used represents a preliminary effort. In this case, the Power Lever Angle (PLA) is compensated and altitude is increased to reduce temporary high damage increment rate. The rationale behind keyboard interference is to intentionally generate some events to verify the DFSA model, control action and simulation response for our convenience. The C++ wrapper acts as a mediator between the simulation (client) and the C++ based controller (server). It is responsible for initiating and maintaining the communication channel, and communicating messages between client and server.

In the implementation of server side, with our perception of the engine operation scenario, an open-loop DFSA model of the plant is created. An Event Generator is implemented to generate the necessary events with the sensor data and other information to drive the state transitions of the DFSA plant model. Supervisor is implemented to restrict the open-loop language of the plant model to improve not only the life of the engine but also the overall mission behavior. Theoretically mission behavior is evaluated by the language measure of the close-loop plant, and experimentally it is evaluated by the ending state of each experiment after executing large number of missions. The purpose here is to try to avoid the engine halt, which has a highly negative contribution to the mission behavior, and try to complete the mission successfully as much as possible. Aborting some missions has a relatively low negative impact on the mission behavior compared with engine halt. On the other hand, the chance of completing the mission will also be reduced. Then the question is how to make the best trade-off, which will be addressed by supervisor design.

The outputs of the server side are the estimated state, various event flags and throttle input for simulation. These variables are sent in the return message structure for the client computer. Other major inputs for the server side software are state transition table, which gives the information about the DFSA model of the plant dynamics, event cost and controllability tables that are used in the Event Generator.

Supervisory action comes into the picture if, at the current state, there exists a controllable event/events. There are two types of requests, one is for flight parameters change in order to reduce the damage increment rate, and the other is for the mission abortion to avoid the engine halt. In this part, the program gives the pilot a certain amount of time to respond to the request. If it is not responded in time, then the default action is taken. Within this respond period, all other event generation procedures are blocked so that the state cannot change. If flight parameters change approved, then it sends the necessary command to the client and gives the client some time to generate the action. During this stage, the program avoids the creation of same event consecutively by blocking the request.

The plant model dynamics are reflected on the server side by using the state transition table. This table is unique for each DFSA model. The rows of the table correspond to events whereas the columns correspond to states. Whenever an event is generated or a supervisory action is taken the related function is called to make the state transition and go to the next state. To do so, the event should be defined in the current state. If it is defined, a state transaction takes place and the new state is written to the output file in order to be examined for the overall mission behavior. Every marked state has its own characteristic value: negative for bad marked states and positive for good marked states. The magnitude of the characteristic value is assigned by the user based on his/her perception of goodness or badness of that particular state.

The GUI, presented in Figure 7.2, serves as the control panel for the pilot. Some important information such as sensor data, estimated state and events will be displayed on the GUI, and the pilot throttle input and control decision will be passed to server function through this GUI. The three round gauges are combustion temperature, oil pressure and high-pressure turbine speed respectively. The three bar gauges are bearing vibration, damage increment rate and accumulated damage respectively. Other information and buttons includes engine running time, current state, last event detected, previous state, throttle input, communication connection button, engine start button, request accept/decline button.

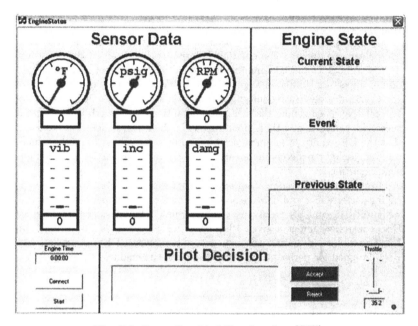

Fig. 7.2. Server Graphical User Interface (GUI)

7.5 Simulation Experiments: Results and Discussion

A series of simulation experiments were designed on the test bed to validate the optimal DES control concept, presented in Section 7.3. The simulation experiments are grouped in four sets, where each set gives different aspects of the DES control concept. The first set is merely a verification experiment, and the second one is on the damage reduction and health management of the aircraft gas turbine engine. The optimal supervisor analysis comes into the picture after conducting the third set of experiments, the identification of event costs. This set of experiments is presented in Section 7.5.1, and is followed by the optimal supervisor synthesis in Section 7.5.2. The last set of experiments, mainly focuses on the performance comparison of the optimal supervisor and the other supervisors that are designed using conventional techniques.

Upon implementation of the software modules on the client and server computers, the first set of simulation experiments has been conducted to verify that functions and communications added on both server and client sides do not affect the simulated engine dynamics. The ensemble of simulation results generates qualitative information on robustness of the DES control system relative to exogenous disturbances and anticipated uncertainties. That is, successful completion of these simulation experiments provides some assurance that the DES control system structure is sufficiently robust to modest uncertainties, including the communication delays, in the dynamics of the engine simulation model.

The second set of simulation experiments was conducted to compare the engine performance and damage accumulation for the unsupervised plant (i.e., without DES control) plant and the supervised plant (i.e., under a typical DES controller) under a given PLA input whose profile is shown in Figure 7.3. Controller No. 2 was selected, out of the several DES controllers that were designed for performance enhancement of the engine.

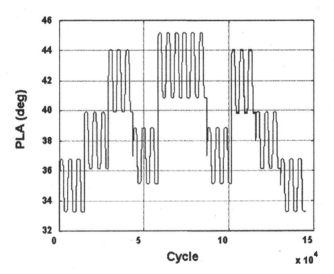

Fig. 7.3. PLA input for the simulation

Under the same fixed PLA input, Figures 7.4 and 7.5 show the engine outputs for the unsupervised plant and supervised plant, respectively. Figure 7.6 shows $\sim 35\%$ damage reduction under DES control.

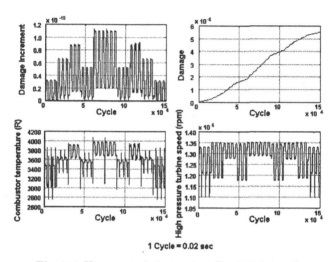

Fig. 7.4. Unsupervised plant output (fixed PLA input)

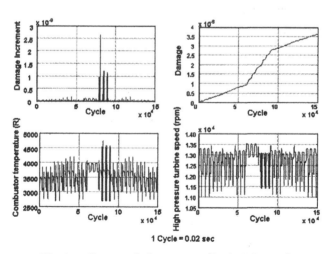

Fig. 7.5. Supervised plant output (fixed PLA input)

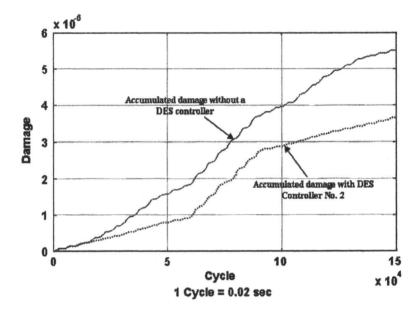

Fig. 7.6. Comparison of Accumulated Damages (fixed inputs)

Although the given throttle inputs are the same, the supervisor modifies them only if high damage increment is detected. When the supervisor detects a high damage increment, i.e. the event generator creates a high damage increment rate event, the supervisor sends a discrete message to the client for the reduction of the PLA and increment of altitude. Another module, the action generator, in the client changes the PLA and altitude in the continuous domain. Consequently, input to the engine simulation is adjusted to reduce the damage increment rate for the supervised case.

Damage increment rate is formulated as a function of high-pressure turbine gas inlet temperature and shaft speed, and accumulated damage is the base of comparison for supervised and unsupervised cases. System health indicator used for this experiments is the efficiency of high pressure turbine that changes in a very slow time scale (e.g., in the order of ten of hours). Thus, it is not possible to see any change in the slow time scale that the experiments are conducted.

Since the pilot inputs are expected to be mostly random throttle excitation to the engine, the last simulation experiment was repeated with random inputs. For (statistically identical) random throttle inputs, Figures 7.7 and 7.8 show the engine outputs for the unsupervised plant and supervised plant, respectively. Comparison of plots in Figure 7.9 indicates ~ 60% damage reduction under DES control.

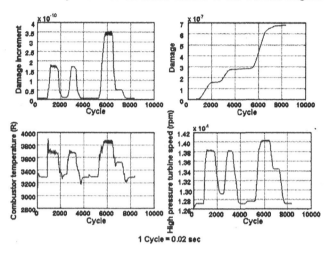

Fig. 7.7. Unsupervised plant output (random PLA input)

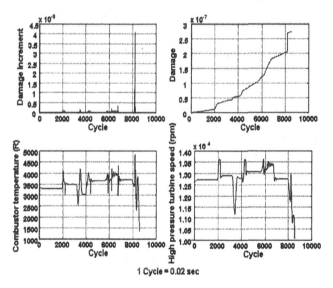

Fig. 7.8. Supervised plant output (random PLA input)

7.5.1 Language Measure Parameter Identification

Analysis and synthesis of an optimal DES controller require the identification of the event cost matrix. Analogous to continuously varying dynamical systems (CVDS), we use techniques of system identification to identify the

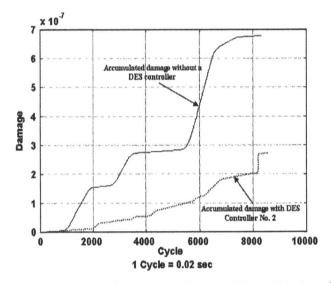

Fig. 7.9. Comparison of Accumulated Damages (pilot random inputs)

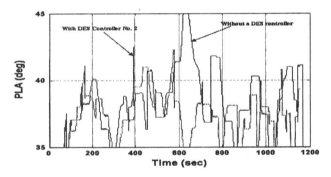

Fig. 7.10. Random input comparison - PLA

language measure parameters of the DFSA plant model - the elements $\tilde{\pi}_{ij}$ of the event cost matrix $\tilde{\Pi}$ (see Definition 7.2.4 in section 7.2). As the number of experiments increases, the identified event costs tend to converge in the Cauchy sense. For stationary operation of the engine, since conditional probabilities of the events can be assumed to be time-invariant, the identified event costs and their uncertainty bounds can be determined. Wang et al. [8] have reported details of the identification procedure and its experimental validation on a robotic test bed. As a typical case, Figure 7.16 presents identification of event costs at state 6 (engine operation under normal damage increment).

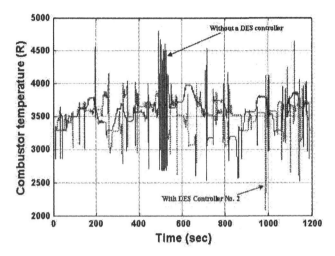

Fig. 7.11. Random input comparison - Combustor Temperature

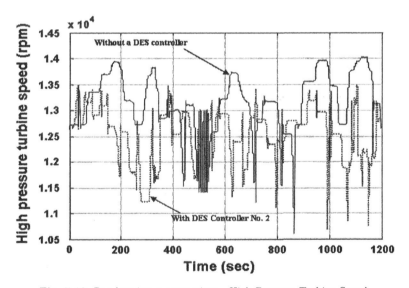

Fig. 7.12. Random input comparison - High Pressure Turbine Speed

The frequency probabilities of state visit and event trigger were monitored and plotted for 100 (statistically) similar simulation experiments.

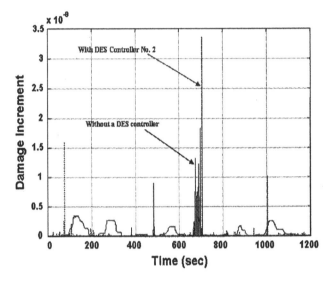

Fig. 7.13. Random input comparison - Damage Increment Rate

Figures 7.10 through 7.15 show the results of simulation experiments for random inputs over a time period at the sampling frequency of 50 hertz (i.e.,

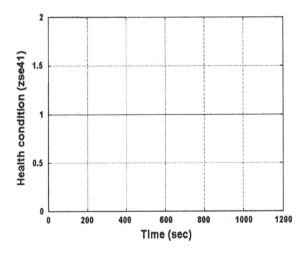

Fig. 7.14. Random input comparison - zse41

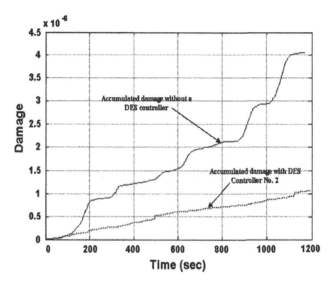

Fig. 7.15. Random input comparison - Accumulated Damage

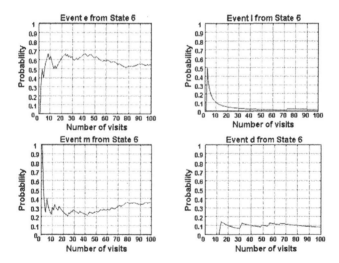

Fig. 7.16. Convergence of event cost identification

a new input is entered at every 20 milliseconds). In each off these figures, the plots in solid line indicate the engine response without a DES controller and those in dotted lines correspond to the engine response under the DES Controller No. 2. In general, DES Controller No. 2 reduces structural damage in engine components. For example, the estimated damage accumulation in the hot sections of the engine is reduced by approximately 75 percent.

7.5.2 Optimal DES Controller Synthesis

The state transition cost matrix Π is determined from the event cost matrix $\tilde{\Pi}$, and the transition function δ of the finite state automaton. Given the state transition cost matrix Π and the state characteristic vector $\bar{\chi}$, the optimal DES controller can be synthesized. The characteristic values of the four marked states in Table 7.1 are assigned as: -0.05, -0.20, -1.00, and +0.20, respectively. These values are assigned based on the designer's perception of the importance of terminating on specific marked states. For example, the bad marked state *Unexpected engine halt* in Table 7.1 is assigned the characteristic value of -1.00 because the single engine aircraft will most likely be destroyed if the DFSA terminates on this state. On the other hand, the good marked state *Mission successful* is assigned the characteristic value of +0.20 based on its relative importance to the loss of the aircraft.

Table 7.2 lists the iterations of optimal control synthesis for the first 16 states. The performance measure of the unsupervised plant is negative at the states 3, 4, 5, 7, 8, 9, 11, 12, 13, 14, and 16 as indicated by bold script in 7.2. All controllable events leading to these states are disabled and the resulting performance measure at Iteration 1 shows sign change at states 3, 4, 5, and 14 as indicated by italics in Table 7.2. All controllable events leading to these states are now re-enabled for further increase in performance as seen in the column under Iteration 2, where sign change occurs only at state 8. Re-enabling all controllable events leads to state 8 increases the performance even further. The synthesis is complete in Iteration 3 (i.e., there is no need to go for the Iteration 4) because there is no sign change from Iteration 2 to Iteration 3. However, the Iteration 4 in Table 7.2 is shown to exhibit that there is no further improvement in the language measure.

The performance of the optimal controller was compared with that of Controller1 and Controller2, which are designed using the conventional procedure [4] [9]. The optimal controller not only yields the best mission performance of all controllers and unsupervised plant, but also reduces the accumulated damage of the unsupervised plant. However, it may not necessarily yield less damage accumulation than all other controllers because damage criteria were not addressed in the formulation of the optimal control policy. Figure 7.17 shows the engine outputs for the input given in Figure 7.3 under the supervision of optimal controller whose damage accumulation is less than that of unsupervised plant, but slightly exceeds that of Controller2. Table 7.3 compares damage accumulation for each case.

Theoretical performance of the supervisors can be associated with the language measure of each supervisor. The language measure of the unsupervised plant and that of the three controllers are listed in Table 7.4.

Table 7.2. Optimal Controller Synthesis Iterations

	Unsup. plant	Iteration 1	Iteration 2	Iteration 3	Iteration 4
State 1	0.1392	0.2396	0.2654	0.2749	0.2749
State 2	0.1406	0.2420	0.2681	0.2777	0.2777
State 3	-0.1826	*0.0475*	0.0762	0.0819	0.0819
State 4	-0.1011	*0.0163*	0.0462	0.0619	0.0619
State 5	-0.4348	*0.0000*	0.0301	0.0346	0.0346
State 6	0.1576	0.2585	0.2833	0.2930	0.2930
State 7	-0.3373	-0.0322	-0.0083	-0.0045	-0.0045
State 8	-0.1134	-0.0077	*0.0126*	0.0268	0.0268
State 9	-0.8250	-0.7116	-0.7061	-0.7050	-0.7050
State 10	0.1249	0.2250	0.2493	0.2590	0.2590
State 11	-0.3759	-0.1857	-0.1706	-0.1665	-0.1665
State 12	-0.1318	-0.0241	-0.0066	-0.0007	-0.0007
State 13	-0.8638	-0.8545	-0.8520	-0.8512	-0.8512
State 14	-0.0622	*0.0372*	0.0628	0.0721	0.0721
State 15	0.3378	0.4372	0.4628	0.4721	0.4721
State 16	-1.0000	-1.0000	-1.0000	-1.0000	-1.0000

The performance of the null and the three supervisors are compared based on observations of mission execution on the simulation test bed. For each con-

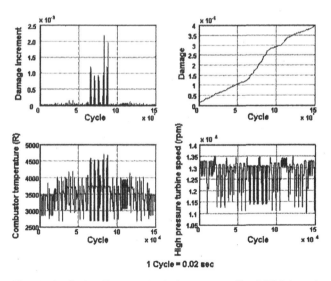

1 Cycle = 0.02 sec

Fig. 7.17. Optimally supervised plant output (fixed PLA input)

Table 7.3. Damage under Different Supervisors

Unsupervised	Controller1	Controller2	OptimalController
5.51×10^{-6}	5.51×10^{-6}	3.65×10^{-6}	3.96×10^{-6}

Table 7.4. Performance under Different Supervisors

$\mu_{Unsupervised}$	$\mu_{Controller1}$	$\mu_{Controller2}$	$\mu_{OptimalController}$
0.1392	0.1312	0.2269	0.2749

troller, 100 missions were simulated, and the mission outcomes were recorded with respect to the characteristic values assigned to the four marked states. Assigning the characteristic values (χ): -0.2, 0.2, -1.0, -0.05 to states: *Mission Abortion off-Ground* (14), *Mission Success* (15), *Engine Halt* (16) and *Mission Abortion on-Ground* (42), respectively, simulated performance of the unsupervised plant and each of the three controllers is calculated as given below:

$\nu_{Unsupervised} = 33 \times (-0.2) + 59 \times 0.2 + 4 \times (-0.05) + 4 \times (-1.0) = 1.00$

$\nu_{Controller1} = 81 \times (0.2) + 19 \times (-1.0) = -2.80$

$\nu_{Controller2} = 25 \times (-0.2) + 68 \times (0.2) + 3 \times (-0.05) + 4 \times (-1.0) = 4.45$

$\nu_{OptimalController} = 31 \times (-0.2) + 63 \times (0.2) + 6 \times (-0.05) + 1 \times (-1.0) = 5.10$

The bar chart in Figure 7.18 shows a comparison of mission behavior for each supervisor under simulation experiments. It is seen that the theoretical performance of the supervisors, listed in Table 7.4, is in qualitative agreement with the experimental results.

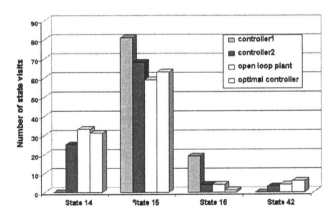

Fig. 7.18. Simulated Performance of Controllers

7.6 Summary and Conclusions

This chapter presents a quantitative approach to the synthesis of an optimal discrete-event supervisory (DES) control of a complex engineering system based on the recent theoretical work in this field [7] [3] [6] [8].

The optimal DES control law has been validated on a gas turbine engine simulation test bed. The plant model in the simulation test bed is built upon the model of a generic turbofan gas turbine engine. The software architecture of the simulation test bed is flexible to adapt arbitrary DFSA models and controller designs and to fit other complex systems such as power plants or robots.

The results of simulation experiments show the discrete-event supervisor is capable of simultaneously reducing structural damage and improving the mission behavior of the engine system. Simulation experiments on the test bed establish feasibility of the optimal DES control theory for applications to large-scale engineering systems. To the best of the authors' knowledge, this is the first application of optimal DES control to a large-scale engineering system reported in open literature.

References

1. J. Fu, C.M. Lagoa, and A. Ray, *Robust optimal control of regular languages with event cost uncertainties*, Proceedings of IEEE Conference on Decision and Control, December 2003, pp. 3209–3214.
2. J. Fu, A. Ray, and C.M. Lagoa, *Optimal control of regular languages with event disabling cost*, Proceedings of American Control Conference, Denver, Colorado, June 2003, pp. 1691–1695.
3. J. Fu, A. Ray, and C.M. Lagoa, *Unconstrained optimal control of regular languages*, Automatica **40** (2004), no. 4, 639–648.
4. P.J. Ramadge and W.M. Wonham, *Supervisory control of a class of discrete event processes*, SIAM J. Control and Optimization **25** (1987), no. 1, 206–230.
5. A. Ray and S. Phoha, *Signed real measure of regular languages for discrete-event automata*, Int. J. Control **76** (2003), no. 18, 1800–1808.
6. A. Surana and A. Ray, *Signed real measure of regular languages*, Demonstratio Mathematica **37** (2004), no. 2, 485–503.
7. X. Wang and A. Ray, *A language measure for performance evaluation of discrete-event supervisory control systems*, Applied Mathematical Modelling **28** (2004), no. 9, 817–833.
8. X. Wang, A. Ray, and A. Khatkhate, *On-line identification of language measure parameters for discrete event supervisory control*, Proceedings of 42nd IEEE Conference on Decision and Control (Maui, Hawaii), December 2003, pp. 6307–6312.
9. X. Wang, A. Ray, S. Phoha, and J. Liu, *J-des: A graphical interactive package for analysis and synthesis of discrete event systems*, Proceedings of American Control Conference (Denver, Colorado), June 2003, pp. 3405–3410.

8

Supervisory Control of Software Systems

Vir Phoha[1], Amit Nadgar[2], Asok Ray[3], and Shashi Phoha[4]

[1] Louisiana Tech University, Ruston, LA 71270 phoha@latech.edu
[2] Louisiana Tech University, Ruston, LA 71270 aun001@latech.edu
[3] The Pennsylvania State University, University Park, PA 16802 axr2@psu.edu
[4] The Pennsylvania State University, University Park, PA 16802 sxp26@psu.edu

Summary. This chapter presents a new paradigm to control software systems based on the Supervisory Control Theory (SCT). The proposed method uses SCT to model the execution process of a software application by restricting the actions of the OS with little or no modifications in the underlying OS. This approach can be generalized to other software applications as the interactions of an application with the Operating System (OS) are modelled at the *process* level as a Deterministic Finite State Automaton (DFSA), called as the "plant". A "supervisor" that controls the plant is also a DFSA that represents a set of control specifications. The supervisor operates synchronously with the plant to restrict the language accepted by the plant to satisfy the control specifications. As a proof-of-concept for software fault management, two supervisors have been implemented under the Redhat Linux 7.2 OS to mitigate overflow and segmentation faults in five different programs. The performance of the unsupervised plant and that of the supervised plant are quantified by using the Language Measure, described in Chapter 1.

Key words: Discrete Event Supervisory Control, Software Systems, Software Fault Management, Language Measure

8.1 Introduction

A computer program is a discrete-event system in which the supervisory control theory (SCT) [15] can be applied to augment a general-purpose operating system (OS) to control and direct a wide range of software applications. This chapter presents a novel SCT-based technique, built upon the formal language theory, to model and control software systems without any structural modifications in the underlying OS. In this setting, the user has the privilege to override the OS actions to control a software application.

SCT is a well-studied paradigm and has been used in a variety of applications. However, for the sake of completeness, this chapter very briefly reviews certain relevant applications of SCT to software systems. Self-adaptation in software systems, where supervisory control is augmented with an adaptive component, is reported in [11]. SCT has been used in the Workflow management paradigm to schedule concurrent tasks through scheduling controllers [20] and also for protocol converters to ensure consistent communications in heterogeneous network environments. Hong et al. [7] has adopted supervisor-based closed-loop control to facilitate software rejuvenation [10] – a technique to improve software reliability during its operational phase.

Ramadge and Wonham [15] present a novel SCT approach to control discrete event systems (DES) using a feedback control mechanism. Here, the DES to be controlled is modeled by the plant automaton $\mathcal{G}$. The uncontrolled behavior of the plant is modified by a supervisor S, such that behavior of the plant is restricted to a subset of $L(\mathcal{G})$. The feedback control that is achieved using the supervisor satisfies a given set of specifications interpreted as sublanguages of $L(\mathcal{G})$ representing legal behavior for the controlled system. Following the approach, this chapter models interactions of software applications with the OS as a deterministic finite state automaton (DFSA) (i.e., a representation for the class of regular language) [9] and applies the SCT for development of a recognizer of this language to control and mitigate faults in software execution. Specifically, the discrete-event supervisor restricts the legal language of the model in an attempt to mitigate the normally detrimental consequences of faults or undesirable events.

The proposed approach first enumerates the events and states of the plant model $\mathcal{G}$. The specifications to control (restrict) the behavior of a computer program by controlling the interactions with the OS, are represented by another DFSA S that has the same event alphabet as the plant model. The parallel combination of S and $\mathcal{G}$ gives rise to a DFSA $(S/\mathcal{G})$, which is the generator under supervisory control [15].

The significant features of the software management approach, proposed in this chapter, are: (i) a novel technique for fault mitigation in software systems; (ii) real-time control of software systems; (iii) runtime behavioral modification and control of the OS with insignificant changes in the underlying OS; (iv) modelling of the OS-Application interactions as symbols (events) in the formal language setting; and (v) accommodation of multiple control policies by varying the state transitions.

Several researchers (see for example [1],[5],[8]) have reported monitoring of programs at run time using automata and in one case *transforming the sequence when it deviates from the specified policy* [1]. However, the approach in [1],[5],[8] is significantly different from that taken in this chapter, where novel principles of supervisory control theory have been used as an extension of Ramadge and Wonham's work [15]. This concept results in parallel, synchronized operation of two or more automata, namely, a plant (i.e., the application or the computer program) and a controller (or a set of controllers), which implements the supervisory control policy on the entire plant rather than on selected components.

The chapter is organized in nine sections and two appendices. Section 8.2 briefly reviews the Supervisory Control Theory. Section 8.3 presents discrete-event modelling of interactions of a computer execution process and the Operating system (OS). Modelling of two supervisors based on a given specification is shown in Section 8.4. Section 8.5 describes an implementation of the supervisory control system for process execution under the Red Hat Linux 7.2. Section 8.6 reviews the Language Measure Theory [21] [16] [19] and a procedure for estimation of the event cost $\tilde{\Pi}$–matrix parameters is presented in Section 8.7. A description of the experiments and the corresponding results are given in Section 8.8. The chapter is concluded in Section 8.9. Appendix A presents the definitions pertinent to Supervisory Control Theory. Appendix B presents a lower bound on the number of experimental observations for identification of the language measure parameters.

8.2 Background

This section reviews the supervisory control theory (SCT) as applied to Discrete Event Systems [15] [3] [4]. A discrete event system (DES) is a dynamical system which evolves due to asynchronous occurrences of certain discrete changes called *events*. A DES has discrete states which correspond to some continua in the evolution of a task. The state transitions in such systems are produced at asynchronous discrete instants of time in response to events and represent discrete changes in the task evolution.

The SCT introduced by Ramadge and Wonham [15] is based on automata and formal language models. Under these models the focus is on the order in which the events occur. A plant is assumed to be the generator of these events. The behavior of the plant model describes event trajectories over the (finite) event alphabet Σ. These event trajectories can be thought of as strings over Σ. Then $L \subseteq \Sigma^*$ represents the set of those event strings that describe the behavior of the plant. In this formal language setting, the concepts of plant and supervisor are discussed in the following sub-sections.

8.2.1 Plant Model

The plant $\mathcal{G}$ is a generator of all the strings in L and is described as a quintuple deterministic finite state automaton (DFSA):

$$\mathcal{G} = (Q, \Sigma, \delta, q_0, Q_m) \tag{8.1}$$

where Q is the set of states for the system with $|Q| = n$; q_0 is the initial state; Σ is the (finite) alphabet of events causing the state transitions; Σ^* is the set of all finite-length strings of events including the empty string ϵ. The state transition function is defined as: $\delta : Q \times \Sigma \to Q$ and $\hat{\delta} : Q \times \Sigma^* \to Q$ is an extension of δ which can be defined recursively as follows :
For any $q \in Q$, $\hat{\delta}(q\epsilon) = q$; for any $s \in \Sigma^*$, $a \in \Sigma$ and $q \in Q$, $\hat{\delta}(q, sa) = \delta(\hat{\delta}(q, s), a)$; $Q_m \subseteq Q$ is the subset of states called the marked (or accepted) states. A marked state represents the completion of a task or a set of tasks. To give this system a means of control, the event alphabet Σ is classified into two categories: **uncontrollable events** ($\sigma \in \Sigma_{uc}$) which can be observed but cannot be prevented from occurring and **controllable events** ($\sigma \in \Sigma_c$) which can be prevented from occurring.

8.2.2 Supervisor

A supervisor S is realized as a function $S = (\mathcal{S}, \phi)$. $\mathcal{S}$ is given by the DFSA quintuple:

$$\mathcal{S} = (X, \Sigma, \xi, x_0, X_m) \tag{8.2}$$

where X is the state set, Σ is the event alphabet which is same as that of the plant, $\xi : \Sigma \times X \to X$ is the transition function and $X_m \subseteq X$ is a subset of marked states; ϕ is a function that maps the states of the supervisor into control patterns $\gamma \in \Gamma$ where $\Gamma = \{0,1\}^{\Sigma_c}$ is the set of all binary assignments to the elements of Σ_c. Each state of the supervisor corresponds to a fixed control pattern where some controllable events are enabled or disabled. Thus the plant is controlled by the supervisor by switching to control patterns corresponding to the supervisor's state of operation which is fully-synchronized [6] with that of the plant.

The methods developed in this chapter use the SCT terminology; interested readers may refer to definitions given in Appendix A.

8.2.3 Supervisory Controller Synthesis

The objective of SCT is to synthesize a supervisor in such a way that the *supervised plant* behaves in accordance with constraints pertaining to its restricted behavior. The control specifications provide the constraints to enforce the restricted behavior of the plant. The following steps delineate the synthesis of a supervisory controller (see Section 8.4 for an illustration of this process; see Appendix A for the definitions of terms in italics referred to in the following steps).

- Model the unsupervised, i.e., open loop physical plant as a DFSA $\mathcal{G}$.
- Provide the specifications of the constrained behavior of $\mathcal{G}$ as English statements. Let K be the formal language obtained for these specifications. Design another DFSA say S with the same event alphabet Σ, to produces the language K.
- Perform the *completion* on S of the specification language K to obtain the automaton $\overline{S}$ containing the dump state.
- Perform *synchronous composition* $(\mathcal{G} \| \overline{S})$.
- The result of the previous operation is used to verify if the specification given by the language $K \subseteq \Sigma^*$ is *controllable*. The *controllability check* [15] [3] is done to ascertain the existence of a supervisor for the given specification.
- If language K is not controllable then it is necessary to determine the *supremal controllable sublanguage* $K^{\uparrow C}$. The resultant automaton for this language is a desired supervisory controller that satisfies the control specification.

The following sections illustrate the synthesis of two supervisors for the OS-process interactions.

8.3 Modeling Operating System - Process Interactions

A process is a program in execution. The terms, process (i.e., a computer program in execution) and software application, are used interchangeably hereafter. For the supervisory controller synthesis (see Section 8.2.3), the OS-process interactions are modelled as a DFSA plant model that is to be controlled by another DFSA, known as the supervisor DFSA.

8.3.1 Plant Model of OS - Process Interactions

The states in the DFSA model represent operational states of a process while the arcs illustrate the system events leading to transitions between these operational states as shown in Figure 8.1. This plant model DFSA is given as a 5-tuple $\mathcal{G} = (Q, \Sigma, \delta, q_1, Q_m)$ with initial state as q_1. Here, $Q = \{q_1, \ldots, q_7\}$, $\Sigma = \{\sigma_1, \ldots, \sigma_{10}\}$, and $Q_m = Q$ is the set of marked states that represent the completion of important operational states of an application program from an OS perspective (see Figure 8.1). Note that $L(\mathcal{G})$ contains only the *legal* (physically admissible) strings that can be achieved starting from the initial state q_1. The plant automaton model $\mathcal{G}$ is made trim (see Definition A.6) so that all states can be reached from the initial state and the automaton is nonblocking. By Definitions A.1 and A.2 in Appendix A, the language $L(\mathcal{G})$ and marked language $L_m(\mathcal{G}) \subseteq L(\mathcal{G})$ of the DFSA $\mathcal{G}$ are derived as:

$$L(\mathcal{G}) = \{s \in \Sigma^* | \delta(q_1, s) \in Q\} \subseteq \Sigma^*$$
$$L_m(\mathcal{G}) = \{s \in \Sigma^* | \delta(q_1, s) \in Q_m\} \subseteq L(\mathcal{G}) \subseteq \Sigma^*,$$

where $L_m(\mathcal{G})$ contains all event strings that terminate on a marked state. By Definition A.3, it can be seen that $L(\mathcal{G})$ is prefix-closed i.e., $L(\mathcal{G}) = pr(L(\mathcal{G}))$.

In this chapter, all events in Σ that are used to model the automaton $\mathcal{G}$ are assumed to be observable [3] events, i.e., the events are visible to the supervisor. These events are constructed by observing *signals* received by a process [2] and by monitoring the free physical memory resource available to the system. As stated in Section 8.2, the event alphabet Σ is partitioned into subsets of controllable events Σ_c and uncontrollable events Σ_{uc}, such that $\Sigma_{uc} \cup \Sigma_c = \Sigma$ and $\Sigma_{uc} \cap \Sigma_c = \emptyset$. A supervisor controls the computer process by selectively disabling the controllable events based on the control specifications. For a given DFSA plant model $\mathcal{G}$, the control system designer usually defines the control specifications that, in turn, generates the supervisor DFSA S.

The states and events of a process $\mathcal{G}$ (with initial state q_1) are listed in Table 8.2 and Table 8.1, respectively. Figure 8.1 presents the state transition diagram for the DFSA $\mathcal{G}$ that captures the behavior of a program in execution as it interacts with the OS. Each of the circles with labels represents a state of the DFSA $\mathcal{G}$. Labels on the arcs are the events contained in the alphabet Σ of the automaton $\mathcal{G}$.

The initial state q_1 in Figure 8.1 is the idle state in which the OS is ready to execute a new process. The program start (event σ_1) is the only

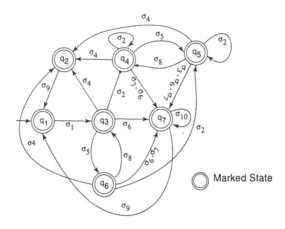

Fig. 8.1. Process-OS interactions modelled by the DFSA $\mathcal{G}$

Table 8.1. List of Events in DFSA $\mathcal{G}$ [$C\equiv$ Controllable; $UC\equiv$ Uncontrollable]

Event	Event Description	Type
σ_1	Start a program	C
σ_2	CPU generated exception due to program error, non-critical hardware failure.	UC
σ_3	Process terminated by OS.	C
σ_4	Execution completed.	UC
σ_5	Low on resources.	UC
σ_6	Process terminated (by an external agent.)	UC
σ_7	Process halted due to lack of resources.	C
σ_8	Resources available.	C
σ_9	Process cleanup upon termination	UC
σ_{10}	Process still resident.	UC

Table 8.2. List of States Q in DFSA $\mathcal{G}$

State	State Description
q_1	Idle state (Ready to execute)
q_2	Execution Completed
q_3	Normal Execution
q_4	Process Fault Detected
q_5	Possible faulty execution as well as low on resources
q_6	Normal execution but low on resources (DP)
q_7	Process Halted

event that can produce a transition from q_1. The system is in state q_3 under normal execution of the program. There are four possible transitions from q_3 in the unsupervised model. The process can remain in q_3 during its life time if it does not incur any exceptions and then move to q_2 by exiting on its own volition using the system call exit() or when the program control flow reaches the last statement of the main procedure main(). However, if the program causes the CPU to generate an exception due to a program error (event σ_2), its state changes to q_4 where the OS executes its exception handler in the context of the process and then notifies the process by a *signal* of the anomalous condition. For a fatal exception, the OS takes the default action of terminating the process even if a *signal handler* [2] is provided. This behavior is modelled as an event σ_3 arising from state q_3. For a non-fatal exception the signal handler (if provided) is executed and the process (if not terminated by the OS) will continue to run at state q_4.

At state q_3, it is possible that the process is terminated by the user which is shown by the event σ_6 that causes a transition to state q_7. The user's decision to terminate the process cannot be controlled and therefore the event σ_6 is made uncontrollable. Such an event is possible from other states of the task and hence σ_6 may occur at all states except q_1. The self-loop of event σ_{10} indicates that the OS has not yet recovered all the resources locked by the process. Event σ_9 represents the actions of the OS to release any resources owned by the process. These resources include memory, open files, and possibly synchronization resources such as semaphores. Any process that completes its execution via event σ_4 traverses to q_2 and thereafter to state q_1 via σ_9.

States q_5 and q_6 depict the scenario in which the system is low on resources such as memory. While q_6 can be reached from q_3 via the low resource event σ_5, q_5 can be reached from q_4 and q_6 via events σ_5 and σ_2, respectively. By access to state q_5, it is shown that a process may incur exceptions when running with inadequate resources. Observe the self-loop at q_4 due to the program error event σ_2 that implies that a task can execute in state q_4 while causing exceptions.

When the system is low on resources (e.g., memory leaks are one of the reasons causing the free physical memory resource to reduce over time and are observed as the running software ages [7]), the OS may take actions to create resources for normal operation of the task and this is accomplished by using the event σ_8. It is also possible that the OS does not take any such action and the task is allowed to execute in the same state where there are chances that it can cause exceptions as shown by the σ_2 self-loop at q_5. The OS may halt a task if it has insufficient resources for further execution in which case σ_7 causes transition to q_7. As long as the process resources remain locked, the system remains at q_7 due to the self-loop σ_{10}. Resumption of the process from state q_7 is not considered. Therefore, as and when the OS releases these resources after it has removed the process's descriptor from memory, event σ_9 captures this behavior by producing a transition to the idle state q_1.

8.3.2 Marked States and Weight Assignments

The purpose of making a state "marked" is that the state should represent completed or important operational phases of the physical plant represented by the model. All states are made marked (i.e., $Q_m = Q$) in this plant model because of their importance during the execution of the process. The marked language of the plant initialized at the state q_1 is given by $L_m(\mathcal{G}) = \{s \in \Sigma^* | \hat{\delta}(q_1, s) \in Q_m\}$. In order to obtain a quantitative measure of $L_m(\mathcal{G})$, The set Q_m of marked states is partitioned into subsets of good and bad marked states, Q_m^+ and Q_m^-, respectively. A good marked state is assigned a positive value while a bad marked state is given a negative value. Each marked state is characterized by assigning a signed real value on the scale of $[-1,1]$ based on the designer's perception of the state's impact on the performance of the software system. A lower value assigned to a marked state implies a greater degradation in the performance of the system in that state.

Marked states of the plant $\mathcal{G}$ are assigned the weights given by the characteristic vector $\overline{\chi_{\mathcal{G}}} = [0.3\ 0.8\ 0.8\ 0.1\ 0.08\ 0.4\ -0.8]^T$. For example, in state q_7, it is assumed that the software application has failed and cannot recover. Therefore, state q_7 is assigned a high negative weight of -0.8 but not -1 as it does not necessarily represent a failure of the OS and other applications. State q_1 has a weight of 0.3 since the application has yet to enter the existing workload on the system; so, q_1 is assigned a relatively lower weight. Together states q_2 and q_3 represent the best desired operation of the system where a process assigned to the CPU for execution, performs its operation and relinquishes control upon completion without producing conditions that adversely affect the system performance; hence, they have weights of 0.8 each. Software faults at state q_4 may cause the software to produce erroneous results, and thus may cause performance degradation in the system, it is assigned a comparatively lower weight of 0.1. In state q_5 the system is in a condition that is relatively worse than that in state q_4 due to depletion of resources (e.g., free physical memory); this is another reason for degraded execution. Therefore, state q_5 has a weight of 0.08 which is lower than that of state q_4. With no prior faulty operation except for reduced resource levels in state q_6, the system is considered to be in an operational state that is better than being in states q_4 or q_5. So, state q_6 is assigned a weight of 0.4 which is higher than the weights of both q_4 and q_5.

8.4 Formulation of Supervisory Control Policies

This section devises two control policies under supervisors S_1 and S_2. The specifications of these two control policies are as follows:

Policy 1. (S_1) – Prevent termination of process on first occurrence of a fatal exception. Enable termination of process on the second occurrence of this exception. Prevent halting of process due to low resources.

Policy 2. (S_2) – Prevent termination of process on first and second occurrence of a fatal exception. Enable termination of process on the
third occurrence of this exception. Prevent halting of process due to
low resources.

Let K_1 and K_2 be the regular languages induced by the specifications of
the supervisors S_1 and S_2 that are synthesized separately in the following two
subsections. The languages K_1 and K_2 are regular and are therefore converted
into the respective supervisor's DFSA [3] [9] as seen in Figures 2(a) and 4(a).
The dotted lines signify that the associated controllable events are disabled
according to the specifications. Note that the event alphabet Σ is common
to the plant model and both supervisors. In addition, it is assumed that all
events are observable (see Section 8.3). However, there are no restrictions on
the state set X (see Equation 8.2) to be same as the state set Q of the plant
model. Table 8.3 lists the states of the supervisory control automata generated
from the specification languages K_1 and K_2:

Table 8.3. List of States in DFSAs S_1 and S_2

S_1's States	S_2's States	State Description
x_1	x_1	Idle state (Ready to execute)
x_2	x_2	Execution Completed
x_3	x_3	Normal Execution
x_4	x_4, x_5	Process Fault Detected. Permit further execution
x_5	x_9	Process Fault Detected. Permit process termination due to fault
x_8	x_8, x_{10}	Low on resources. Permit execution inspite of faults
x_9	x_{11}	Low on resources. Permit process termination due to fault
x_6	x_6	Normal execution but low on resources (DP)
x_7	x_7	Process Halted

8.4.1 Synthesis of Supervisor S_1

Let S_1 be the DFSA (see Figure 2(a)) created for the specification given by
Policy 1. Let K_1 be the language of this supervisor DFSA. Figure 2(a) shows
that there are additional states in the supervisor automaton than in plant
model automaton $\mathcal{G}$.

The supervisor DFSA in Figure 2(a) is analyzed to examine the restrictions enforced on the plant's behavior by the specification S_1 (Policy 1). The
supervisor captures the start of a process through event σ_1 which produces a
transition from state x_1 to x_3. The first time a process causes an exception the
supervisor observes it through the transition from state x_3 to state x_4. The

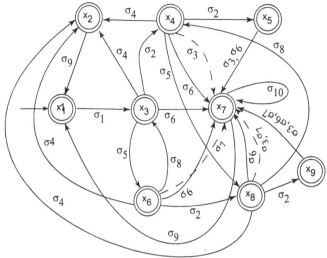

(a) DFSA of the control specification language K_1 for supervisor S_1 obtained from the specification given by Policy 1.

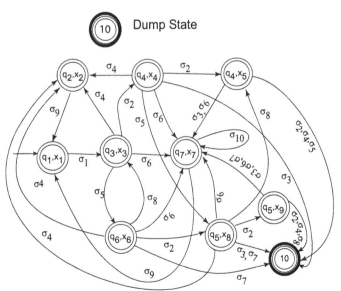

(b) DFSA of $(\mathcal{G}\|\overline{S_1})$ where $\overline{S_1}$ is the *Completion* of automaton S_1 given in Figure 2(a)

Fig. 8.2. Figures (a) and (b) show the DFSA of the sublanguage K_1 for supervisor S_1 and its failure to satisfy the *Controllability Condition* respectively. Dashed lines show controllable events disabled at states as per the specification given by Policy 1.

process is not terminated immediately as the event σ_3 at state x_4 is disabled to prevent the OS from terminating the process.

State q_4 in Figure 8.1 shows a self-loop due to the additional occurrences of exceptions while running in that state. The specification of S_1 restricts the language of $\mathcal{G}$ such that an event sequence it produces contains at the most two instances of the event σ_2. Therefore, state x_5 is created to eliminate the effects of the self-loop; consequently, the supervisor moves on the second occurrence of an exception, i.e., the event σ_2. Note that x_4 and x_5 are states where the process is not low on a resource such as memory.

Similar counting of exceptions is performed when the process shows a degraded state of operation due to less than required resources. The supervisor has state x_6 to indicate that the process is running on low physical memory. As modeled in $\mathcal{G}$, under low resource conditions the process moves to state q_5 when an exception occurs and continues there if additional exceptions occur (shown by the self-loop at state q_5 in Figure 8.1). Under this situation the supervisor is required to restrict the language produced by the process and so states x_8 and x_9 are added in the supervisor automaton. The policy states that the process should not halt due to less available memory; consequently, the event σ_7 is disabled at states x_6 and x_8. Enabling σ_7 at state x_9 does not affect the specification as the OS terminates the process when event σ_3 is enabled at state x_9.

It is necessary to test the *controllability condition* (see Definition A.9, [15]) on K_1 to determine whether the specification language K_1 is controllable. To this end, the DFSA $(\mathcal{G}\|\overline{S_1})$ (see Figure 2(b)) is generated by synchronous composition (see Definition A.7 in Appendix A) of $\mathcal{G}$ and $\overline{S_1}$, where $\overline{S_1}$ is the *completion* of the automaton S_1 in Figure 2(a), with dump state. In Figure 2(b) state 10 is the *dump state*.

It is noted that $K_1 \subseteq L(\mathcal{G})$ is controllable if and only if for each $s \in pr(K_1)$ and $u \in \Sigma_{uc}$ such that $su \in L(\mathcal{G})$ and $su \in pr(K_1)$. Figure 2(b) shows that there are uncontrollable events from states (q_4, x_5) and (q_5, x_9) leading to the dump state 10, and event strings such as $(\sigma_1\sigma_5\sigma_8\sigma_2\sigma_2\sigma_5) \notin pr(K_1)$. Therefore, the controllability condition on K_1 is violated and the controller K_1 is not controllable.

Since the controllability condition does not hold on K_1, one needs to determine the *supremal controllable sublanguage* (see Definition A.10 and [3]) of K_1. The strings with uncontrollable prefixes (e.g., $(\sigma_1\sigma_5\sigma_8\sigma_2\sigma_2)$) are removed to obtain the supremal controllable sublanguage $K_1^{\uparrow C}$ of K_1. It suffices to eliminate the states reached by their execution in $(\mathcal{G}\|\overline{S_1})$ to remove the set of strings with uncontrollable prefixes. Consequently, states (q_4, x_5), (q_5, x_9) and the *dump state* 10 are eliminated from $(\mathcal{G}\|\overline{S_1})$ as seen in Figure 8.3. The DFSA in Figure 8.3 is the supervised plant $(S_1/\mathcal{G})$ that satisfies the control policy (i.e., Policy 1) in the supervisor S_1.

The weights assigned to the marked states of the supervised plant automaton $S_1/\mathcal{G}$ are given by the characteristic vector:

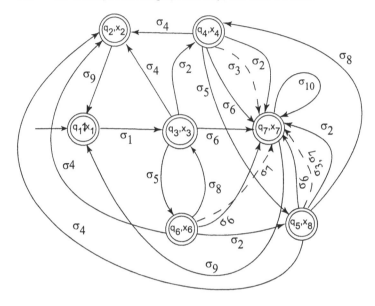

Fig. 8.3. DFSA for Supervised process $(S_1/\mathcal{G})$ using the supremal controllable sublanguage $K_1^{\uparrow C}$ for supervisor S_1 induced by the specification Policy 1. Dashed lines show the controllable events that are disabled at the states.

$$\overline{\chi}_{S_1/\mathcal{G}} = [0.3\ 0.8\ 0.8\ 0.1\ 0.08\ 0.4\ -0.8]^T$$

The same rationale is applied for assigning weights similar to what is done for the unsupervised plant $\mathcal{G}$ in Section 8.3.2.

8.4.2 Synthesis of Supervisor S_2

Let S_2 be the supervisor DFSA for the specification given by Policy 2 based on the language K_2 as shown in Figure 4(a). The supervisor S_2 is similar to the supervisor S_1 with the difference that it allows the process to execute even after the second occurrence of an exception.

The supervisor S_2 is differs from supervisor S_1 in the number of exceptions it permits before the OS is allowed to terminate the application. Similar to supervisor S_1, counting is made possible by adding states . Figure 4(a) shows that states x_5 and x_9 are added to represent the scenario when the system is not low on resources; and states x_{10}, and x_{11} are added to satisfy the control policy.

Similar to the supervisor S_1 (see Section 8.4.1) the supervisor S_2 also observes the first time a process causes an exception through the transition from

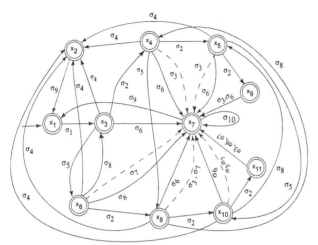

(a) DFSA of the control specification K_2 for supervisor S_2 obtained from the specification given by Policy 2

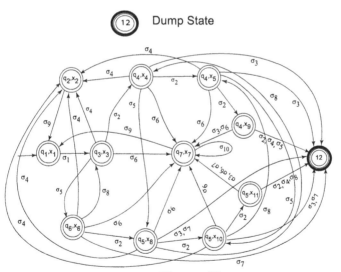

(b) DFSA of the *Completion* $\mathcal{G}\|\overline{S_2}$, where $\overline{S_2}$ is the *Completion* of S_2

Fig. 8.4. Figures (a) and (b) show the DFSA of the sublanguage K_2 for supervisor S_2 and its failure to satisfy the *Controllability Condition* respectively. Dashed lines show controllable events disabled at states as per the specification given by Policy 2.

state x_3 to state x_4 on event σ_2. Therefore the exception counting mechanism is similar to that of the supervisor S_1, only here the language of $\mathcal{G}$ is restricted such that an event trajectory contains at the most three instances of the event σ_2. Therefore, states x_5 and x_9 are added. Note that x_4, x_5 and x_9 are the states where the process does not suffer from inadequacy of the memory resources. Disabling the OS's action of terminating the process is shown in Figure 4(a) by disabling the controllable event σ_3 at states x_4, x_5 as well as at states x_8 and x_{10}. Figure 4(b) shows that specification K_2 as there are uncontrollable events leading to the dump state. Therefore, the supremal controllable sublanguage $K_2^{\uparrow C}$ is generated similar to what was done for supervisor S_1. Figure 8.5 shows the automaton for the desired controller $(S_2/\mathcal{G})$ that satisfies Policy 2.

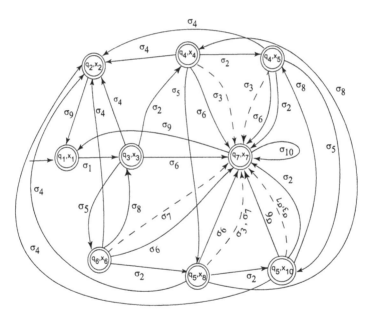

Fig. 8.5. DFSA for Supervised process $S_2/\mathcal{G}$ using the supremal controllable sublanguage $K_2^{\uparrow C}$ for supervisor S_2 given by the specification of Policy 2. Dashed lines show the controllable events that are disabled at the states.

The weights assigned to the marked states of the supervisor $S_2/\mathcal{G}$ are given by the characteristic vector:

$$\overline{X}_{S_2/\mathcal{G}} = [0.3\ 0.8\ 0.8\ 0.1\ 0.1\ 0.08\ 0.08\ 0.4\ -0.8]^T$$

Similar rationale, as of the unsupervised plant $\mathcal{G}$, is applied for the weight assignment here. New states that are created are assigned weights depending on the state of the plant with which the supervisor state has combined. Therefore, states (q_4, x_5) and (q_5, x_{10}) are assigned weights of 0.1 and 0.08 respectively.

8.5 Implementation of Supervisory Control

This section presents the proposed framework of the software architecture for realizing control in software systems. Figure 8.6 shows the supervisory control system comprised of 4 separate components in labeled ellipses above the rectangular box and represent individual user level processes. However only a combination of the three fundamental processes P_{sensor}, E_{gen}, $P_{supervisor}$ exercise control over the OS.

The objective here is to achieve control over an application indirectly by being able to access the OS. The process, denoted by the ellipse labeled P_1, is indirectly controlled by the supervisor. The functional processes P_{sensor}, E_{gen}, $P_{supervisor}$ in Figure 8.6 are implemented as user level processes with the capability to insert *kernel modules* [17] so that the control path can be changed to satisfy the supervisory control policy. Entire implementation of the supervisory control system is under the Red Hat Linux 7.2 OS [2]. For simplicity of illustration, this chapter shows the implementation for a single process under supervisory control.

8.5.1 Components of the SCT Framework

The components of the supervisory control system are listed below: (i)P_{sensor} - The sensor/actuator system has two functions: (a) Monitoring signals and collection of resource status information received from the OS; and (b) implementation of supervisory decisions. (ii)E_{gen} - The event generator maps the resource measurements and signals received by the process into corresponding higher-level events. (iii)$P_{supervisor}$ - The supervisor automaton is designed based on the control specification.

The Sensor/Actuator System

The sensor system implements two threads to track a process. The threads perform (i) signal capturing; and (ii) process information collection, respectively. To intercept the signals received by the process the sensor uses `ptrace()` and `wait()` [2] system calls . In this application, the Linux */proc* file system (indexed by the process identifier) has been used to obtain relevant system information of process P_1. This information pertains to the computer process parameters (e.g., code size and memory usage) and is sent as a different message type to the event generator. The event generator uses this data for the generation of high level events.

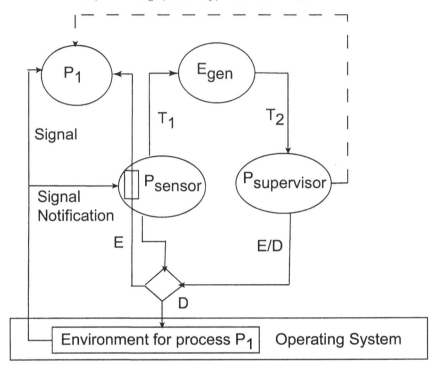

Fig. 8.6. Framework for supervisory control of OS

The Event Generator

The supervisor process that executes the controlled language automaton $S/\mathcal{G}$ can perceive only high level events (see Table 8.1). E_{gen} maps sensor signals and the data provided by the process information collection thread as higher level events compatible with $P_{supervisor}$. E_{gen} implements both one-one mapping as well as the many-one mapping of signals to events. Mapping of signals and resource information to high level events is shown in Table 8.4.

The Supervisor Automaton

The controlled language automaton (e.g., $S_1/\mathcal{G}$ and $S_2/\mathcal{G}$) is executed by the supervisor process. The supervisor is implemented to control a process using the information provided by E_{gen} as a message containing the event through the communication link T_2. The supervisor dynamically makes decisions of disabling certain controllable events depending on its current state. The status of each controllable event – *enabled* (E) or *disabled* (D) is then sent to P_{sensor}

Table 8.4. Mapping signals and data collected from /proc filesystem into events of the plant $\mathcal{G}$. TM = Total Physical Memory, FM = Available Free Physical Memory

Event	Mapping Conditions
σ_1	SIGSTOP sent to process when sensor uses the ptrace() system call.
σ_2	wait() system call's WSTOPSIG macro returns the signal numbers SIGFPE or SIGSEGV.
σ_3	wait() system call's WIFSIGNALED macro returns true but termination not due to SIGKILL signal (WTERMSIG $\neq$ SIGKILL).
σ_4	wait() system call's WIFEXITED macro evaluates to a value > 0.
σ_5	Free memory available is below user defined threshold (FM $<$ 5% of PM).
σ_6	SIGKILL detected by wait() system call's WTERMSIG macro.
σ_7	Free memory available below user defined threshold (FM $<$ 1% of PM).
σ_8	Available free memory above user defined threshold (FM $>$ 5% of PM).
σ_9	/proc contains no entry of process ID (PID).
σ_{10}	/proc contains process information after SIGKILL or WTERMSIG $\neq$ SIGKILL.

via the protocol E/D. The sensor process also performs the task of an actuator to enforce supervisor's decisions that may cause the actions of the OS to change. In the end, the results are reflected in the actions taken by the OS on the application program tracked by the sensor.

8.5.2 Operation of SCT Framework

RedHat Linux 7.2 is a multitasking operating system. Here a kernel maintains a status repository for each process. The kernel keeps track of information such as whether a particular process is currently executing on the CPU or whether it is blocked on an event, the address space assigned to the process, and the files the process has opened during its life time, etc. To facilitate the maintenance of this information a process descriptor is implemented in the kernel using the data structure task_struct. Figure 8.6 shows a rectangular box labeled environment of the process P_1 which represents the information base for this process.

In this application, BSD sockets [18] have been used to communicate between different components of the system. Different communication special-purpose protocols that have been developed (see Figure 8.6) are described below:

Protocol T_1 shown by the uni-directional arrow from P_{sensor} to E_{gen}. Protocol T_2 shown by the uni-directional arrow from E_{gen} to $P_{supervisor}$. Protocol E/D shown by the arrow going from $P_{supervisor}$ to P_{sensor}. P_{sensor} uses the protocol T_1 to send the raw signal information it collects to E_{gen}. The event generator creates events from these raw signals and then using T_2 sends this information to the supervisor process. The supervisor process executing the $S/\mathcal{G}$ automaton is only capable of disabling or enabling the events. Depending on the current state of the automaton $S/\mathcal{G}$, the E/D (*enable/disable*) decision is sent by the $P_{supervisor}$ to the sensor. Although the sensor has control over the process, the objective here is to control the process by making OS deviate from its normal path.

The sensor process has the ability to dynamically load and unload *kernel modules* [17] to create the deviations from the original control path. These deviations are needed when a control policy designates a particular event to be disabled in a particular state. This action is shown by the arrow labelled D emanating from the decision box. For an enabled event, the process is made aware of the situation as indicated by the arrow labelled E going to the process. Figure 8.6 also shows a set of arrows emanating from the environment of the process. The labels *Signal* and *Signal Notification* relates directly to the mechanism the ptrace() system call uses for making the sensor process aware of a signal (e.g. SIGSTOP, SIGWINCH, SIGFPE, SIGSEGV etc.) a process has received.

8.6 Performance Measurement of Controllers

In general, the controlled sublanguages of a plant language could be different under different controllers that satisfy their respective policies. The partially ordered set of sublanguages can then be totally ordered using their respective performance measures [21] [16] [19]. This total ordering using performance measures facilitates comparison between the unsupervised and the supervised plant models. To this end, the language measure theory is applied to compute the performance measures of the DFSAs. The plant performance under different supervisors is quantitatively evaluated and compared by making use of the language measure, described in detail in Chapter 1. The results of performance comparison are presented in Section 8.8.2. A brief review of the language measure is presented below for completeness of this chapter.

8.6.1 Language Measure Theory

This section briefly reviews the concept of signed real measure of regular languages [21] [16] [19]. Let the discrete-event behavior of a plant be modelled as a DFSA $G_i \equiv (Q, \Sigma, \delta, q_i, Q_m)$ as given in Equation (8.1) and therefore Q, δ and Q_m have the same meaning. Let the cardinality of the event alphabet be $|\Sigma| = m$, and the dump state [15] if any, be excluded so that $|Q| = n$. The languages generated by the plant DFSA when initialized at a state q_i are are obtained definitions A.1 and A.2 and are given below.

Definition 8.6.1 *A DFSA G_i, initialized at $q_i \in Q$, generates the language $L(G_i) \equiv \{s \in \Sigma^* : \delta^*(q_i, s) \in Q\}$ and its marked sublanguage $L_m(G_i) \equiv \{s \in \Sigma^* : \delta^*(q_i, s) \in Q_m\}$.*

The language $L(G_i)$ is partitioned as the non-marked and the marked languages, $L^o(G_i) \equiv L(G_i) - L_m(G_i)$ and $L_m(G_i)$, consisting of event strings that, starting from $q \in Q$, terminate at one of the non-marked states in $Q - Q_m$ and one of the marked states in Q_m, respectively. The set Q_m is partitioned into Q_m^+ and Q_m^-, where Q_m^+ contains all *good* marked states that are desired to terminate on and Q_m^- contains all *bad* marked states that are desired not to terminate on, although it may not always be possible to avoid the bad states while attempting to reach the good states. The marked language $L_m(G_i)$ is further partitioned into $L_m^+(G_i)$ and $L_m^-(G_i)$ consisting of good and bad strings that, starting from q_i, terminate on Q_m^+ and Q_m^-, respectively.

A signed real measure $\mu : 2^{\Sigma^*} \to \Re \equiv (-\infty, \infty)$ is constructed for quantitative evaluation of every event string $s \in \Sigma^*$. The language $L(G_i)$ is decomposed into null (i.e., $L^o(G_i)$), positive (i.e., $L_m^+(G_i)$), and negative (i.e., $L_m^-(G_i)$) sublanguages.

Definition 8.6.2 : *The language of all strings that, starting at a state $q_i \in Q$, terminates on a state $q_j \in Q$, is denoted as $L(q_i, q_j)$. That is, $L(q_i, q_j) \equiv \{s \in L(G_i) : \delta^*(q_i, s) = q_j\}$.*

Definition 8.6.3 : *The characteristic function that assigns a signed real weight to state-partitioned sublanguages $L(q_i, q_j), i = 0, 1, \ldots, n - 1$, is defined as: $\chi : Q \to [-1, \ 1]$*
    such that

$$\chi(q_j) \in \begin{cases} [-1, \ 0) \ if \ q_j \in Q_m^- \\ \{0\} \quad if \ q_j \notin Q_m \\ (0, \ 1] \ if \ q_j \in Q_m^+ \end{cases}$$

Definition 8.6.4 : *The event cost is conditioned on a DFSA state at which the event is generated, and is defined as $\tilde{\pi} : \Sigma^* \times Q \to [0, 1]$ such that $\forall q_j \in Q, \forall \sigma_k \in \Sigma, \forall s \in \Sigma^*, \ \tilde{\pi}(\sigma_k, q_j) = 0$ if $\delta(q_j, \sigma_k)$ is undefined; $\tilde{\pi}(\varepsilon, q_j) = 1; \ \tilde{\pi}(\sigma_k, q_j) \equiv \tilde{\pi}_{jk} \in [0, 1) \ ; \quad \sum_k \tilde{\pi}_{jk} < 1; \ \tilde{\pi}(\sigma_k s, q_j) = \tilde{\pi}(\sigma_k, q_j) \tilde{\pi}(s, \delta(q_j, \sigma_k)).$*

The event cost matrix, denoted as $\tilde{\Pi}$–matrix, is defined as:

$$\tilde{\Pi} = \begin{bmatrix} \tilde{\pi}_{00} & \tilde{\pi}_{01} & \cdots & \tilde{\pi}_{0m-1} \\ \tilde{\pi}_{10} & \tilde{\pi}_{11} & \cdots & \tilde{\pi}_{1m-1} \\ \vdots & & \ddots & \vdots \\ \tilde{\pi}_{n-10} & \tilde{\pi}_{n1} & \cdots & \tilde{\pi}_{n-1m-1} \end{bmatrix}$$

Now the measure of a sublanguage of the plant language $L(G_i)$ is defined in terms of the signed characteristic function χ and the non-negative event cost $\tilde{\pi}$.

Definition 8.6.5 *Given a DFSA $G_i \equiv \langle Q, \Sigma, \delta, q_i, Q_m \rangle$ the cost ν of a sublanguage $K \subseteq L(G_i)$ is defined as the sum of the event cost $\tilde{\pi}$ of individual strings belonging to K $\nu(K) = \sum_{s \in K} \tilde{\pi}(s, q_i)$*

Definition 8.6.6 *The signed real measure μ of a singleton string set $\{s\} \subset L(q_i, q_j) \subseteq L(G_i) \in 2^{\Sigma^*}$ is defined as: $\mu(\{s\}) \equiv \chi(q_j) \tilde{\pi}(s, q_i) \quad \forall s \in L(q_i, q_j)$.*

The signed real measure of $L(q_i, q_j)$ is defined as: $\mu\left(L(q_i, q_j)\right) \equiv \sum\limits_{s \in L(q_i, q_j)} \mu\left(\{s\}\right)$
and the signed real measure of a DFSA G_i, initialized at the state $q_i \in Q$, is denoted as: $\mu_i \equiv \mu(L(G)) = \sum_j \mu\left(L(q_i, q_j)\right)$

Definition 8.6.7 *The state transition cost of the DFSA is defined as a function* $\pi : Q \times Q \to [0, 1)$ *such that* $\forall q_j,\ q_k \in Q,\ \ \pi(q_j, q_k) = \sum\limits_{\sigma \in \Sigma :\, \delta(q_j,\, \sigma) = q_k} \tilde{\pi}(\sigma,\, q_j) \equiv$
π_{jk} *and* $\pi_{jk} = 0$ *if* $\{\sigma \in \Sigma :\ \delta(q_j,\, \sigma) = q_k\} = \emptyset$. *The state transition cost matrix, denoted as Π —matrix, is defined as:*

$$\Pi = \begin{bmatrix} \pi_{00} & \pi_{01} & \cdots & \pi_{0n-1} \\ \pi_{10} & \pi_{11} & \cdots & \pi_{1n-1} \\ \vdots & & \ddots & \vdots \\ \pi_{n-10} & \pi_{n-11} & \cdots & \pi_{n-1n-1} \end{bmatrix}$$

Wang and Ray [21] have shown that the measure $\mu_i \equiv \mu(L(G_i))$ of the language $L(G_i)$, with the initial state q_i, can be expressed as: $\mu_i = \sum_j \pi_{ij} \mu_j + \chi_i$ where $\chi_i \equiv \chi(q_i)$. Equivalently, in vector notation: $\bar{\mu} = \Pi\bar{\mu} + \bar{\chi}$ where the measure vector $\bar{\mu} \equiv [\mu_1\ \mu_2\ \cdots\ \mu_n]^T$ and the characteristic vector $\bar{\chi} \equiv [\chi_1\ \chi_2\ \cdots\ \chi_n]^T$.

8.6.2 Probabilistic Interpretation

The signed real measure (Definition 6) for a DFSA is based on the assignment of the characteristic vector and the event cost matrix. As stated earlier, the characteristic function is chosen by the designer based on his/her perception of the states' impact on system performance. On the other hand, the event cost is an intrinsic property of the plant. The event cost $\tilde{\pi}_{jk}$ is conceptually similar to the state-based conditional probability as in Markov Chains, except for the fact that it is not allowed to satisfy the equality condition $\sum_k \tilde{\pi}_{jk} = 1$. Note that $\sum_k \tilde{\pi}_{jk} < 1$ is a sufficiency condition for convergence of the language measure [21] [22].

With this interpretation of event cost, $\tilde{\pi}[s, q_i]$ (Definition 4) denotes the probability of occurrence of the event string s in the plant model G_i starting at state q_i and terminating at state $\hat{\delta}(s, q_i)$. Hence, $\nu(L(q, q_i))$ (Definition 5), which is a non-negative real number, is directly related (but not necessarily equal) to the total probability that state q_i would be reached as the plant operates. The language measure $\mu_i \equiv \mu(L(G_i)) = \sum_{q \in Q} \mu(L(q_i, q)) = \sum_{q \in Q} \nu(L(q_i, q))\chi(q)$ is then directly related (but not necessarily equal) to the expected value of the characteristic function. Therefore, in the setting of language measure, a supervisor's performance is superior if the supervised plant is more likely to terminate at a *good* marked state and/or less likely to terminate at a *bad* marked state.

8.7 Estimation of Event Cost $\tilde{\Pi}$–Matrix Parameters

The theoretical bound, N_b, on the number of experimental observations (see equation 8.6 in Appendix B)is used to compute the event cost matrix parameters for a given pair of parameters δ and ε. Let p_{ij} denote the transition probability of event σ_j on the state q_i, i.e.,

$$p_{ij} = \begin{cases} P[\sigma_j|q_i], & \text{if } \exists q \in Q, \ s.t. \ q = \delta(q_i, \sigma_j) \\ 0, & \text{otherwise} \end{cases}$$

and its estimate $\hat{p}_{ij}$ that is obtained from the ensemble of experiments and/or simulation data. Let the indicator function $I_t(i,j)$ represent the occurrence of event σ_j at time t if the system was in state q_i at time $t-1$ where t represents a generalized time epoch, for example, the experiment number. Formally, $I_t(i,j)$ is expressed as:

$$I_t(i,j) = \begin{cases} 1, & \text{if } \sigma_j \text{ is observed at state } q_i \\ 0, & \text{otherwise} \end{cases}$$

Let $N_t(i)$, denoting the number of incidents of reaching the state q_i up to the time instant t, be a random process mapping the time interval up to the instant t into the set of nonnegative integers. Similarly, let $n_t(i,j)$ denote the number of occurrences of the event σ_j at the state q_i up to the time instant t.

The plant model, in general, is an inexact representation of the physical plant, which is assumed to manifest itself either as unmodelled events that may occur at each state. or as unaccounted states in the model. Let Σ_i^u denote the set of unmodeled events at state i of the DFSA. Therefore, the residue $\theta_i = 1 - \sum_j \tilde{\pi}_{ij}$ can be viewed as the probability of the all the unmodelled events, emanating from state q_i. Let $\Sigma^u \equiv \cup_i \Sigma_i^u \ \forall i \in \mathcal{I} \equiv \{1, \ldots, n\}$, at each state q_i, $P[\Sigma^u|q_i] = \theta_i \in (0,1)$ and $\sum_i \tilde{\pi}_{ij} = 1 - \theta_i$. Therefore, an estimate of the $(i,j)^{th}$ element in $\tilde{\Pi}$–matrix, denoted by $\hat{\tilde{\pi}}_{ij}$, is obtained as:

$$\hat{\tilde{\pi}}_{ij} = \hat{p}_{ij}(1 - \theta_i) \tag{8.3}$$

Since $\theta_i \ll 1$, an alternative approximation approach is taken, for ease of implementation, by setting $\theta_i = \theta \ \forall i$. The parameter $0 < \theta \ll 1$ is selected from the numerical perspective, based on the fact that the sup-norm $\|\mu\|_\infty \leq \theta^{-1}$ [21] [19].

The algorithm for estimation of the event cost $\tilde{\Pi}$ matrix parameters is presented below.

Algorithm: Event Cost Matrix Parameter Estimation

(1) Initialize $\forall q_i \in Q \ n_0(i) = 0$ and $N_0(i) = 0$
(2) Compute N for a given δ and ε using equation 8.6 (see Appendix B)

```
/*For each state qi ∈ Q check if it occurs in the tth experiment.
```

```
    If a state q_i occurs then increment its occurrence count.
    Similarly if an event σ_j occurs at state q_i in the t^th experiment
    then increment the event occurrence count for event σ_j at state
    q_i.
    In order to obtain stable state transition probabilities this
    loop is repeated until all states reach the upper bound computed
    in step 2.
    */
```

(3) do

$$\text{for } i=1 \text{ to } |Q|$$
$$N_t(i) = N_{t-1}(i) + I_t(i)$$
$$n_t(i,j) = n_{t-1}(i,j) + I_t(i,j)$$
$$\text{end}$$
$$\text{until } \forall\ q_i \in Q,\ \min\{N_t(i)\} \le N$$

```
    /*Each element π̂̃_ij of the Π̃--matrix is an estimate of the true
```
transition probability p_{ij} such that $\hat{p}_{ij} = \frac{n_t(i,j)}{N_t(i)}$ and $\lim_{N_t(i) \to \infty} \hat{p}^t_{ij} = p_{ij}$.
```
    */
```

(4) for i=1 to $|Q|$
$$\quad\text{for } j=1 \text{ to } |\Sigma|$$
$$\quad\quad \hat{p}^t_{ij} = \frac{n_t(i,j)}{N_t(i)}$$
$$\quad\quad \tilde{\Pi}[i][j] \leftarrow \hat{\tilde{\pi}}_{ij} = \hat{p}^t_{ij}(1-\theta)$$
$$\quad\text{end}$$
$$\text{end}$$

This estimation procedure is conservative since some elements of the $\tilde{\Pi}$–matrix reach the bound before others under the stopping condition of $\forall q_i \in Q,\ \min\{N_t(i)\} \le N$.

8.7.1 $\tilde{\Pi}$–Matrix Parameter Identification by Simulation

The estimation of event cost $\tilde{\Pi}$–matrix parameters (see Section 8.7) takes into consideration the number of occurrences of a state $q_i \in Q$ and the number of occurrences of some event $\sigma_j \in \Sigma$ at state q_i, of a DFSA ($\mathcal{G}$, $S_1/\mathcal{G}$ and $S_2/\mathcal{G}$) under study. Ideally, the parameter estimation algorithm requires real-time field data that provide information on the distribution of types of faults, or when to inject a simulated fault. When a fault is injected, E_{gen} maps the lower level signals into the set of high level events that are known a priori; and thus the model is capable of responding to these events [13].

Simulation Procedure   A uniform random number generator has been used to produce a random number in the range of $[0,1)$ for simulating the production of events. A region in range $[0,1)$ is designated for each state;

these regions are mapped into events defined on that state. In this way, each simulation run is defined when the DFSA starts from the initial state and returns back to the initial state, designated as the idle state (see Tables 8.2 and 8.3). The position of a random number in the range of $[0, 1)$ decides the event to be produced. If the random number lies in a region that maps to a event that is disabled by the control policy, then the event is not produced. The random number generation is continued until a random value that maps to an enabled event is generated. This process is repeated at each state. For each such enabled event, the DFSA makes transitions leading to a simulated path of an application under the influence of software faults. Upon completion of each simulation run, an event trajectory is created along with the event production frequency $(n_t(i,j))$ and the state frequency $(N_t(i))$. For example, a typical observation sequence is: $\sigma_1, \sigma_5, \sigma_2, \sigma_6, \sigma_{10}, \sigma_{10}, \sigma_{10}, \sigma_9$. To account for the occurrence of unmodelled events, the parameter θ (see Section 8.7) is set as: $\theta = 0.02$. Therefore, each row element of the $\tilde{\Pi}$–matrix is multiplied by $(1 - \theta) = 0.98$.

8.8 Experimental Results and Discussion

A prototype of the proposed fault mitigation algorithm is implement as a proof of the concept. The test facility is built upon five different computer programs where each program performs simple mathematical operations in a loop. These programs that are written in the C language were injected with overflow errors (division by zero) and memory errors (e.g., segmentation faults). The following subsections report typical results of controlling these computer programs using the two supervisory control policies S_1 and S_2, described in Section 8.3 and the performance of the unsupervised plant and the supervised plant under the two control policies.

8.8.1 Observations of Real-Time Software Control Under SCT Framework

The default result of both division by zero and segmentation faults are termination of the process. When the divide by zero exception occurs the sensor detects the exception. According to both the control policies the process is not terminated on the very first instance of the exception implying the event σ_3 is disabled. To enforce the disabling decision of the supervisor, the sensor inserted its own handlers at runtime by dynamically loading and unloading *kernel modules* and then made the process aware of the exception. The new handlers incremented the instruction pointer so that same exception does not occur again. When the instruction pointer is incremented the first time, it produces a segmentation fault. The sensor intercepted the segmentation fault and then modified the `task_struct` of the process to accommodate new handlers for segmentation faults. Here too the handlers incremented the instruction pointer. Finally, as the application resumed from the exception handling, it continued with its normal execution. When the number of exceptions incurred by the software application was more than that prescribed by the control language, the supervisor permitted the OS to terminate the process.

8.8.2 Performance Measure: Results and Discussion

The event cost matrix is computed using the simulation data (see Section 8.7.1) for a $\delta = 0.005$, $\varepsilon = 0.075$ and $\theta = 0.02$ by applying the Event Cost Matrix Parameter Estimation procedure (see sections 8.7 and Appendix B). This section computes the state transition cost matrix Π from the event cost $\tilde{\Pi}$- matrix using Definition 8.6.7. Each Π matrix is followed by the $\overline{\chi}$ vector of each DFSA. As stated earlier, the $\overline{\chi}$ matrix of each DFSA contains weights that are assigned to the marked states of the DFSA. For the rationale used in assigning weights to marked states refer to sections 8.3.2, 8.4.1, 8.4.2, 8.6.1 and 8.6.2. The states of the controllers $(S_1/\mathcal{G})$ and $(S_2/\mathcal{G})$ are mapped into row indexes of their corresponding Π matrices. State q_1 of the plant model and (q_1, x_1) of both supervisors are mapped into the 1st row of their respective Π-matrices.

The matrix $\Pi_\mathcal{G}$ is the Π–matrix for the uncontrolled plant model $\mathcal{G}$.

$$
\Pi_\mathcal{G} = \begin{bmatrix}
0.0 & 0.0 & 0.98 & 0.0 & 0.0 & 0.0 & 0.0 \\
0.98 & 0.0 & 0.0 & 0.0 & 0.0 & 0.0 & 0.0 \\
0.0 & 0.2515 & 0.0 & 0.2427 & 0.0 & 0.2466 & 0.2392 \\
0.0 & 0.1997 & 0.0 & 0.2013 & 0.2055 & 0.0 & 0.3736 \\
0.0 & 0.1413 & 0.0 & 0.1505 & 0.1542 & 0.0 & 0.5341 \\
0.0 & 0.1671 & 0.184 & 0.0 & 0.2292 & 0.0 & 0.3997 \\
0.4887 & 0.0 & 0.0 & 0.0 & 0.0 & 0.0 & 0.4913
\end{bmatrix}
$$

The $\overline{\chi}_\mathcal{G}$ vector for the marked states of the DFSA $\mathcal{G}$ are chosen as:

$$
\overline{\chi}_\mathcal{G} = \begin{bmatrix} 0.3 & 0.8 & 0.8 & 0.1 & 0.08 & 0.4 & -0.8 \end{bmatrix}^T
$$

The language measure for the DFSA of the uncontrolled plant model is computed by the vector space formula $\overline{\mu} = [I - \Pi]^{-1}\overline{\chi}$ (see Section 8.6.1)and is given below by matrix $\overline{\mu}_\mathcal{G}$.

$$
\overline{\mu}_\mathcal{G} = \begin{bmatrix} 8.332 & 8.965 & 8.195 & 7.157 & 6.927 & 7.564 & 6.431 \end{bmatrix}^T
$$

While the $\overline{\chi}$–vector (i.e., weights assigned to the marked states) of the DFSA $S_1/\mathcal{G}$ remains the same as that for the unsupervised plant DFSA $\mathcal{G}$, the Π–matrix and $\overline{\mu}$–vector for the supervised plant DFSA $S_1/\mathcal{G}$ were obtained as follows:

$$\Pi_{S_1/\mathcal{G}} = \begin{bmatrix} 0.0 & 0.0 & 0.98 & 0.0 & 0.0 & 0.0 & 0.0 \\ 0.98 & 0.0 & 0.0 & 0.0 & 0.0 & 0.0 & 0.0 \\ 0.0 & 0.2591 & 0.0 & 0.2408 & 0.0 & 0.2383 & 0.2417 \\ 0.0 & 0.2454 & 0.0 & 0.0 & 0.2396 & 0.0 & 0.495 \\ 0.0 & 0.3114 & 0.0 & 0.2308 & 0.0 & 0.0 & 0.4378 \\ 0.0 & 0.2495 & 0.2542 & 0.0 & 0.2325 & 0.0 & 0.2438 \\ 0.4959 & 0.0 & 0.0 & 0.0 & 0.0 & 0.0 & 0.4841 \end{bmatrix}$$

$$\overline{\mu}_{S_1/\mathcal{G}} = \begin{bmatrix} 10.178 & 10.774 & 10.079 & 9.00 & 9.117 & 9.777 & 8.232 \end{bmatrix}^T$$

Similarly, the Π–matrix and $\overline{\mu}$–vector for the supervised plant DFSA $S_2/\mathcal{G}$ were obtained as follows:

$$\Pi_{S_2/\mathcal{G}} = \begin{bmatrix} 0.0 & 0.0 & 0.98 & 0.0 & 0.0 & 0.0 & 0.0 & 0.0 & 0.0 \\ 0.98 & 0.0 & 0.0 & 0.0 & 0.0 & 0.0 & 0.0 & 0.0 & 0.0 \\ 0.0 & 0.2491 & 0.0 & 0.2417 & 0.0 & 0.0 & 0.0 & 0.2475 & 0.2417 \\ 0.0 & 0.2536 & 0.0 & 0.0 & 0.2284 & 0.2618 & 0.0 & 0.0 & 0.2362 \\ 0.0 & 0.2229 & 0.0 & 0.0 & 0.0 & 0.0 & 0.2466 & 0.0 & 0.5105 \\ 0.0 & 0.2569 & 0.0 & 0.2354 & 0.0 & 0.0 & 0.2661 & 0.0 & 0.2216 \\ 0.0 & 0.2092 & 0.0 & 0.0 & 0.2331 & 0.0 & 0.0 & 0.0 & 0.5377 \\ 0.0 & 0.2597 & 0.2491 & 0.0 & 0.0 & 0.2343 & 0.0 & 0.0 & 0.2368 \\ 0.4892 & 0.0 & 0.0 & 0.0 & 0.0 & 0.0 & 0.0 & 0.0 & 0.4908 \end{bmatrix}$$

The $\overline{\chi}_{S_2/\mathcal{G}}$ vector, presented below, shows the choice of weights assigned to the marked states of $S_2/\mathcal{G}$.

$$\overline{\chi}_{S_2/\mathcal{G}} = \begin{bmatrix} 0.3 & 0.8 & 0.8 & 0.1 & 0.1 & 0.08 & 0.08 & 0.4 & -0.8 \end{bmatrix}^T$$

$$\overline{\mu}_{S_2/\mathcal{G}} = \begin{bmatrix} 10.635 & 11.223 & 10.547 & 9.626 & 9.253 & 9.605 & 9.243 & 10.24 & 8.65 \end{bmatrix}^T$$

In each experiment, q_1 is the initial state for $\mathcal{G}$ and (q_1, x_1) is the the initial state for both $(S_1/\mathcal{G})$ and $S_2/\mathcal{G}$. Therefore, 1st element of the measure μ–vector is of interest in each case. The results are: $\mu(L_m(\mathcal{G})) = 8.33$; $\mu(L_m(S_1/\mathcal{G})) = 10.17$; and $\mu(L_m(S_2/\mathcal{G})) = 10.63$. The measure for $S_2/\mathcal{G}$ is larger those of $\mathcal{G}$ and $S_1/\mathcal{G}$. Therefore, the supervisor S_2 provides the required services better than the unsupervised application and the supervisor S_1 under selected software faults. The rationale for better fault tolerance property of S_2 is that the supervised plant $S_2/\mathcal{G}$ disables the events σ_3 and σ_7; in contrast, the application would have been terminated under S_1 at state (q_5, x_8).

8.9 Summary and Conclusions

This chapter presents a language-theoretic technique to model and control software systems without any structural modifications to the application and the underlying operating system. A supervisory controller is designed to control the execution of a software application to mitigate the detrimental effects of faults while the program is in execution. The supervisor directs a software system to a safe state under the occurrence of selected faults or other undesirable events e.g., low resources and physical memory. The process of synthesizing a supervisor is described, starting from natural language (e.g., English language) specifications of the control objectives and the controllability of each specification is demonstrated. The concept of supervisory control applied to software systems is implemented under the Red Hat Linux 7.2 Operating System and the experimental results are presented.

This chapter also describes the procedure for computing the language measure parameters, specifically the event cost $\tilde{\Pi}$–matrix, identified from test data to obtain the performance measures of the supervisors and the uncontrolled plant model. These language measures assure that the performance of the controlled software application is superior to the uncontrolled application. The qualitative analysis using the language measure determines when it is advantageous to run the software application under the control of a given supervisor over another. The control technique augmented with the language measure has a wide applicability to mitigate faults in software systems and to provide a ranking among a family of supervisors.

Appendix A
Definitions and Nomenclature

This appendix reproduces definitions and nomenclature of concepts in language and automata theory from [15] [12] [3], which are generally used in the supervisory control theory.

Definition A.1. *Language generated - $L(\mathcal{G})$*
The language generated by a generator $\mathcal{G}$ in Equation(8.1) is $L(\mathcal{G}) = \{s \in \Sigma^* | \hat{\delta}(q_0, s) \text{ is defined}\}$

Definition A.2. *Language Marked - $L_m(\mathcal{G})$*
The language marked by a generator $\mathcal{G}$ in Equation(8.1) is $L_m(\mathcal{G}) = \{s \in L(\mathcal{G}) | \hat{\delta}(q_0, s) \in Q_m\}$

Definition A.3. *Prefix-closure - $pr(L)$*
Let $L \subseteq \Sigma^*$, then $pr(L) = \{s \in \Sigma^* | \exists t \in \Sigma^* (st \in L)\}$. L is said to be prefix-closed if $L = pr(L)$. In other words $pr(L)$ contains all the prefixes of the language L.

Definition A.4. *Accessibility*
For a DFSA $\mathcal{G}$ given in Equation(8.1), the set of states reachable from a state $p \in Q$ in $\mathcal{G}$ is denoted by $Re_{\mathcal{G}}(p) = \{q \in Q| \exists s \in \Sigma^* \ s.t. \ \hat{\delta}(p, s) = q\}$. $\mathcal{G}$ is said to be accessible if $Re_{\mathcal{G}}(p) = Q$, i.e., if all the states in $\mathcal{G}$ are reachable from the initial state q_0.

Definition A.5. *Co-accessibility*
A DFSA $\mathcal{G}$ given by Equation(8.1) is said to be co-accessible if $\forall q \in Q$, $Re_{\mathcal{G}}(q) \cap Q_m \neq \emptyset$ i.e., at least one marked state is reachable from each state of $\mathcal{G}$.

Definition A.6. *Trimness*
An automaton $\mathcal{G}$ given by Equation(8.1) is said to be trim if it is both accessible and coaccessible.

Definition A.7. *Synchronous Composition*
Synchronous Composition of DFSAs is used to represent the concurrent operation of component systems. Given two DFSAs $M_1 = (Q_1, \Sigma_1, \delta_1, q_{0,1}, Q_{m,1})$ and $M_2 = (Q_2, \Sigma_2, \delta_2, q_{0,2}, Q_{m,2})$, the synchronous composition of M_1 and M_2 is a DFSA defined as follows: $M = M_1\|M_2 = (Q, \Sigma, \delta, q_0, Q_m)$ where $Q = Q_1 \times Q_2$; $\Sigma = \Sigma_1 \cup \Sigma_2$; $q_0 = (q_{0,1}, q_{0,2})$; $Q_m = Q_{m,1} \times Q_{m,2}$; $\forall q = (q_1, q_2) \in Q, \sigma \in \Sigma$ the transition function $\delta(\cdot, \cdot)$ is defined as follows:

$$
\delta(q, \sigma) = \begin{cases} (\delta_1(q_1, \sigma), \delta_2(q_2, \sigma)) & \delta_1(q_1, \sigma), \delta_2(q_2, \sigma) \ defined, \ \sigma \in \Sigma_1 \cap \Sigma_2 \\ (\delta_1(q_1, \sigma), q_2) & \delta_1(q_1, \sigma) \ defined, \ \sigma \in \Sigma_1 - \Sigma_2 \\ (q_1, \delta_2(q_2, \sigma)) & \delta_2(q_2, \sigma) \ defined, \ \sigma \in \Sigma_2 - \Sigma_1 \\ undefined & otherwise \end{cases}
$$

and if $\Sigma_1 = \Sigma_2$, then $L_m(M_1\|M_2) = L_m(M_1) \cap L_m(M_2)$.

Definition A.8. *Completion - $\overline{M}$*
The completion of a DFSA $M = (Y, \Sigma, \alpha, y_0, Y_m)$, given by the DFSA $\overline{M} = (\overline{Y}, \Sigma, \overline{\alpha}, y_0, Y_m)$, where $\overline{Y} = Y \cup \{y_D\}$, with $y_D \notin Y$ (y_D denotes the dump state), and $\forall \overline{y} \in \overline{Y}, \sigma \in \Sigma$

$$
\overline{\alpha}(\overline{y}, \sigma) = \begin{cases} \alpha(\overline{y}, \sigma) \ if \ \overline{y} \in Y, \alpha(\overline{y}, \sigma) \ defined \\ y_D \quad otherwise \end{cases}
$$

Definition A.9. *Controllability*
For an unsupervised plant model $\mathcal{G}$ given by Equation(8.1) let $K \subseteq \Sigma^*$ be a set of specification that restricts the plant's behavior. The language K is said to be controllable with respect to $\mathcal{G}$ and Σ_u if $pr(K)\Sigma_u \cap L(\mathcal{G}) \subseteq pr(K)$. This condition on K is called the controllability condition. The controllability condition is equivalent to saying that the supervisor never disables an uncontrollable event in G, formally $\forall s \in \Sigma^*, \sigma \in \Sigma_u$, if $s \in pr(K)$, $s\sigma \in L(\mathcal{G})$, then $s\sigma \in pr(K)$.

Definition A.10. *Supremal Controllable Sublanguage $K^{\uparrow C}$*
For an unsupervised plant model $\mathcal{G}$ given by the automaton in Equation(8.1)

let $K \subseteq \Sigma^*$ be a set of specification that restricts the plant's behavior. If the language K is not controllable, then the task is to find the "largest" sublanguage of K that is controllable, where "largest" is in terms of inclusion. Let $\mathcal{C}_{in}$ be the class of controllable sublanguages (L') of K where, $\mathcal{C}_{in} = \{L' \subseteq K | pr(L')\Sigma_u \cap L(\mathcal{G}) \subseteq pr(L')\}$ then $K^{\uparrow C} = \bigcup_{L' \in \mathcal{C}_{in}(K)} L'$

Appendix B
Bound on Experimental Observations

This section presents a stopping rule to determine a bound on the number of experiments Stopping rule to be conducted for identification of the $\tilde{\Pi}$-matrix parameters. The objective is to achieve a trade-off between the number of experimental observations and the estimation accuracy. A stopping rule is presented below.

A bound on the required number of samples samples!bound on is estimated using the Gaussian structure for the binomial distribution that is an approximation of the sum of a large number of independent and identically distributed (i.i.d.) Bernoulli trials. Denoting $\tilde{\pi}_{ij}$ as p and $\hat{\tilde{\pi}}_{ij}$ as $\hat{p}$, it follows that $\hat{p} \sim \mathcal{N}(p, \frac{p(1-p)}{N})$, where $E[\hat{p}] = p$ and $\mathrm{Var}[\hat{p}] = \sigma^2 \approx \frac{p(1-p)}{N}$, provided that the number of samples N is sufficiently large. Let $\mathbf{X} \equiv \hat{p} - p$, then $\frac{\mathbf{X}}{\sigma} \sim \mathcal{N}(0, 1)$. Given, $0 < \varepsilon \ll 1$ and $0 < \delta \ll 1$, the problem is to find a bound N_b on the number of experiments such that $P\{|\mathbf{X}| \geq \varepsilon\} \leq \delta$. Equivalently,

$$P\left\{\frac{|\mathbf{X}|}{\sigma} \geq \frac{\varepsilon}{\sigma}\right\} \leq \delta \tag{8.4}$$

that yields a bound on N as:

$$N_b \geq \left(\frac{\theta^{-1}(\delta)}{\varepsilon}\right)^2 p(1 - p) \tag{8.5}$$

where $\theta(x) \equiv 1 - \sqrt{\frac{2}{\pi}} \int_0^x e^{-\frac{t^2}{2}} dt$. Since the parameter p is unknown, the fact that $p(1 - p) \leq 0.25 \ \forall p \in [0, 1]$ is used to obtain a (possibly conservative) estimate of the bound in terms of the specified parameters ε and δ as:

$$N_b \geq \left(\frac{\theta^{-1}(\delta)}{2\varepsilon}\right)^2 \tag{8.6}$$

The above estimate of the bound on the required number of samples, which suffices to satisfy the specified $\varepsilon - \delta$ criterion, is less conservative than that obtained from the Chernoff bound and is significantly less conservative than that obtained from Chebyshev bound [14], which does not require the assumption of any specific distribution of $\mathbf{X}$ except for finiteness of the rth ($r = 2$) moment.

References

1. L. Bauer, J. Ligatti, and D. Walker, *More enforceable security policies.*, Foundations of Computer Security Workshop (2002).
2. D.P. Bovet and M. Cesati, *Understanding the linux kernel*, O'Reilly & Associates, January 2001.

3. C.G. Cassandras and S. Lafortune, *Introducrion to discrete event systems*, Kluwer Academic, 1999.
4. F. Charbonnier, H. Alla, and R. David, *Supervised control of discrete-event dynamic systems*, IEEE Transactions on Control Systems Technology **7** (1989), no. 2, 175–187.
5. U. Erlingsson and F.B. Schneider, *SASI enforcement of security policies: A retrospective.*, New Security Paradigms Workshop (1999), 87–95.
6. M. Heymann, *Concurrency and discrete event control*, IEEE Control Systems Magazine (1990), 103–112.
7. Y. Hong, D. Chen, L. Li, and K. Trivedi, *Closed loop design for software rejuvenation*, SHAMAN - Self-Healing, Adaptive and self-MANaged Systems (2002).
8. F.B. Schneider, *Enforceable security policies*, ACM Transactions on Information and System Security 3(1) (2002), 30-50.
9. J. E. Hopcroft, R. Motwani, and J. D. Ullman, *Introduction to automata theory, languages, and computation, 2nd ed.*, Addison-Wesley, 2001.
10. Y. Huang, C. Kintala, N. Kolettis, and N. Fulton, *Software rejuvenation: analysis, module and applications*, Proceedings of 25th International Symposium on Fault-tolerance Computing. (1995).
11. G. Karsai and A. Ledeczi, *An approach to self adaptive software based on supervisory control*, IWSAS 2001 (Balatonfured, Hungary), 2001.
12. R. Kumar and V. Garg, *Modeling and control of logical discrete event systems*, Kluwer Academic, 1995.
13. V. Phoha, A. Nadgar, A. Ray, J. Fu, and S. Phoha, *Supervisory control of software systems for fault mitigation*, June 2003, pp. 2229–2233.
14. M. Pradhan and P. Dagum, *Optimal monte carlo estimation of belief network inference*, Twelfth Conference on Uncertainty in Artificial Intelligence (Portland, OR), 1996, pp. 446–453.
15. P.J. Ramadge and W.M. Wonham, *Supervisory control of a class of discrete event processes*, SIAM J. Control and Optimization **25** (1987), no. 1, 206–230.
16. A. Ray and S. Phoha, *Signed real measure of regular languages for discrete-event automata*, Int. J. Control **76** (2003), no. 18, 1800–1808.
17. A. Rubini, *Linux device drivers*, O'Reilly & Associates, June 2001.
18. W.R. Stevens, *Unix network programming*, 2 ed., vol. 1, Addison-Wesley Longman, Singapore, 1999.
19. A. Surana and A. Ray, *Signed real measure of regular languages*, Demonstratio Mathematica **37** (2004), no. 2, 485–503.
20. C. Wallace, P. Jensen, and N. Soparkar, *Supervisory control of workflow scheduling*, Proceedings of International Workshop on Advanced Transaction Models and Architectures (Goa), August - September 1996.
21. X. Wang and A. Ray, *A language measure for performance evaluation of discrete-event supervisory control systems*, Applied Mathematical Modelling **28** (2004), no. 9, 817–833.
22. X. Wang, A. Ray, and A. Khatkhate, *On-line identification of language measure parameters for discrete event supervisory control*, Proceedings of 42nd IEEE Conference on Decision and Control (Maui, Hawaii), December 2003, pp. 6307–6312.

9

Supervisory Control of Malicious Executables in Software Processes

Xin Xu[1], Vir V. Phoha[2], Asok Ray[3], and Shashi Phoha[4]

[1] Louisiana Tech University, LA 71272 xxu001@latech.edu
[2] Louisiana Tech University, LA 71272 phoha@latech.edu
[3] The Pennsylvania State University, University Park, PA 16802 axr2@psu.edu
[4] The Pennsylvania State University, University Park, PA 16802 sxp26@psu.edu

Summary. This chapter models the execution of a software process as a discrete event system that can be represented by a Deterministic Finite State Automaton (DFSA) in the discrete event setting. Supervisory Control Theory (SCT) is applied for on-line detection of malicious executables and prevention of their spreading. The language measure theory, described in Chapter 1, is adapted for performance evaluation and comparison of the unsupervised process automaton and five different supervised process automata. Simulation experiments under different scenarios show the rate of correct detection of malicious executables to be 88.75%.

Key words: Discrete Event Supervisory Control, Software Systems, Malicious Executable, Language Measure

9.1 Introduction

With the increasing use of computers, the number of malicious programs and their propagation rate keep increasing too. There are various approaches to detect malicious programs. These approaches are generally classified into two categories: the specific method and the generic method [15]. The specific method is a relatively simple and economical approach to detect viruses, but it is only applicable to known viruses [5][14][29] and requires the anti-virus tool to be updated frequently. Anti-virus tools, which use a specific method, require prior knowledge, such as a unique string or a unique sequence of byte code, of a particular virus before it can be detected. In the generic method, anti-virus tools monitor the behavior of programs or system activities and the users are warned as soon as any suspicious behavior or activity is detected. While the tools of the generic method are capable of detecting unknown viruses, their major drawback is high probability of false alarms. That is, legitimate activities may be frequently flagged as suspicious; this may result in disruption of the normal work or lead the user to ignore the warnings.

This chapter introduces a novel approach to detect and control spreading of viruses, which complements the current technology of virus detection. In the proposed approach, execution of the software process is modelled as a discrete event system, in which the system calls are mapped as events. It makes use of discrete event supervisory (DES) control theory, pioneered by Ramadge and Wonham [22], to prevent the reproduction and propagation of malicious executables. The language measure theory [29] [27], described in detail in Chapter 1, has been used to evaluate the performance of both unsupervised and supervised plant model automata. The language measure provides a quantitative comparison of plant performance under different supervisors.

Novelties and significant contributions of the proposed approach to control of malicious executables are summarized below:

- Online detection and prevention of spreading of malicious executables.
- Reliable and consistent protection of software system and application processes.

- Modelling of system calls as events in the formal language setting.

This chapter reports an extension of earlier work [21] on virus control, where supervisory control theory was applied to file virus as a proof of concept. Only one supervisor was developed and the rate of correct detection was 75%. In contrast, the present method yields a correct detection rate of 88.75% because of more precise definition of events.

The rest of the chapter is organized as follows. Section 9.2 presents the relevant literature on system calls and state transition analysis of software systems. Section 9.3 introduces supervisory control theory and language measure theory. Section 9.4 describes process modelling as a deterministic finite state automaton (DFSA) and the development of supervisory control algorithms to prevent spreading of file viruses after their detection. Section 9.5 presents simulation results and performance analyses of the supervised and unsupervised process automata using the language measure theory. Section 9.6 concludes the chapter along with recommendations for future work.

9.2 Related Research

This section briefly reviews the literature on system calls and state transition analysis of software systems. The system call trace is an ordered sequence of system calls that a process performs during its execution. Many researchers ([2][3][4][6][7][16][20]) have used system call traces as a type of audit data to perform intrusion detection (ID). They distinguish normal and abnormal activities based on the system call traces to detect intrusions. It has been shown that system calls are critical to appropriate usage of the process resources.

Several researchers (e.g.,[1][8][10][9][18][19]) have developed ID systems based on state transition analysis of audit data, but the technique of state transition analysis has been seldom used for virus detection. To the knowledge of authors, only LeCharlier and Swimmer [15] have used state transition diagrams to represent infection scenarios for virus detection in the framework of rule-based expert systems. The rules are generated by collecting and analyzing the computer audit data. In contrast, the virus detection system, proposed in this chapter, is built upon the principle of supervisory control theory in the framework of state transition analysis to detect new malicious executables. An MS-DOS trace tool [26] has been adopted to trace the normal and malicious executables. A state transition model has been developed based on the trace results. The proposed approach belongs to the class of generic methods.

9.3 Background

This section briefly introduces the supervisory control theory [22] and language measure theory [29] [27]. Supervisory control theory is a well-studied paradigm and has been used in many applications. An example is control of software systems [11], where the workflow management paradigm is used to schedule concurrent tasks through scheduling controllers [28]; another example is usage of protocol converters to ensure consistent communications in heterogeneous network environments [13]. The novelty of applying the Language measure theory to detection and prevention of virus spreading is that it allows quantitative evaluation of the unsupervised process and its comparison with the controlled behavior under different supervisory controllers.

9.3.1 Introduction to Supervisory Control Theory

A discrete event system is a dynamical system with asynchronous occurrence of physical events at discrete time instants. Using Ramadge and Wonham framework [22], a discrete event system is modelled as a regular language that can be realized by a deterministic finite state automaton (DFSA), also called the plant model.

The plant behavior is modelled as a DFSA $G_i \equiv (Q, \Sigma, \delta, q_i, Q_m)$, where Q is the set of states with cardinality $|Q| = n$ and q_i is the initial state; Σ is the event alphabet with cardinality $|\Sigma| = \ell$, and the Kleene closure Σ^* is the set of all finite-length strings of events including the empty string ϵ; state transitions are governed by a (possibly partial) function $\delta : Q \times \Sigma \to Q$, and $\hat{\delta} : Q \times \Sigma_* \to Q$ is an extension of δ; and the set of marked states is denoted by $Q_m \subseteq Q$.

The event alphabet Σ is partitioned into the subset of controllable events, Σ_c, and the subset of uncontrollable events, Σ_u. Let $\Gamma \equiv \{0,1\}^{\Sigma_c}$ be the set of all binary assignments to the elements of Σ_c. The function $\gamma : \Sigma_c \to \{0,1\}$ is called a control pattern in the sense that a controllable event $\sigma \in \Sigma_c$ is said to be enabled by γ if $\gamma(\sigma) = 1$ or disabled by γ if $\gamma(\sigma) = 0$.

The objective is to design a supervisor that is capable of switching control patterns in a way that a given discrete event process satisfies certain specified constraints. Formally, a supervisor S is a pair $S = (S, \phi)$, in which $S = (X, \Sigma, \xi, x_1, X_m)$ is a deterministic automaton with the state set X, the alphabet Σ (the same alphabet as that of the plant automaton), transition function $\xi : \Sigma \times X \to \Gamma$, initial state x_1 and marked state set $X_m \subset X$, ϕ is a function that maps the states of the supervisor into control pattern γ. Thus for each $x \in X, \gamma := \phi(x) : \{0,1\}^{\Sigma_c}$.

Control policies are formulated based on the goal of the supervised automaton. Different control specifications may correspond to different γ functions. It is possible to design several supervisors for the same plant model G with different γ functions to achieve the desired behavior of the supervised plant under different mission objectives.

The goal of the discrete event supervisory (DES) (control system, presented in this chapter, is detection and spreading of malicious executables in a process while it is executing. The control policy here is to disable an event that might lead the DFSA to terminate on one of the bad marked states. The key idea is to mitigate the (software) damage caused by malicious executables as much as possible. Several DES controllers may achieve this goal from different perspectives. The role of the language measure is to quantify the performance of the controlled plant behavior so that the best supervisor can be selected.

9.3.2 Language Measure theory

The signed measure of a regular language is capable of quantitative evaluating the controlled behavior of the DFSA under different supervisors. The language measure is constructed based on assignment of an event cost matrix and a characteristic vector. While the details are presented in Chapter 1, this subsection briefly reviews the concept of signed real measure and provides pertinent definitions.

Definition 9.3.1 *A DFSA G_i, initialized at $q_i \in Q$, generates the language:*

$$L(G_i) = \left\{ s \in \Sigma^* : \hat{\delta}(q_i, s) \in Q \right\}$$

and its sublanguage is given as:

$$L_m(G_i) = \left\{ s \in \Sigma^* : \hat{\delta}(q_i, s) \in Q_m \right\}$$

The language $L(G_i)$ is partitioned as the non-marked $L^o(G_i)$ consisting of strings starting from $q_i \in Q$ and terminating at $q_j \in Q_m$. The set Q_m is partitioned into Q_m^+ and Q_m^-, where Q_m^+ contains all good marked states that are desired to terminate on and Q_m^- contains all bad marked states that are desired to avoid, although it may not always be possible to avoid the bad states while attempting to reach the good states. The marked language $L_m(G_i)$ is further partitioned into $L_m^+(G_i)$ and $L_m^-(G_i)$ consisting of good and bad strings that, starting from q_i, terminate on q_j that belongs to Q_m^+ and Q_m^-, respectively.

Definition 9.3.2 *Let $L(q_i, q_j)$ denote the set of all strings that, starting at q_i,*
terminate on a state q_j for $\forall q_i, q_j \in Q$, i.e., $L(q_i, q_j) = \left\{ s \in L(G_i) | \hat{\delta}(q_i, s) = q_j \right\}$.

Definition 9.3.3 *The characteristic function that assigns a signed real weight to state partitioned sublanguages $L(q_i, q_j)$, $i = 1, 2, \ldots, n$, is defined as:*
$\chi : Q \to [-1, 1]$ *such that*

$$\chi(q_j) \in \begin{cases} [-1, 0) & \text{if } q_j \in Q_m^- \\ \{0\} & \text{if } q_j \in Q - Q_m \\ (0, 1] & \text{if } q_j \in Q_m^+ \end{cases}$$

The assigned values are chosen based on the designer's perception of the states impact on the system performance. In the vector notation, the set of characteristic values is written as: $X = [\chi_1 \chi_2 \cdots \chi_n]^T$, which is called the characteristic vector.

Definition 9.3.4 *The event cost is defined as follows:*
$\tilde{\pi} : \Sigma^* \times Q \to [0, 1)$ *such that* $\forall q_j \in Q, \forall \sigma_k \in \Sigma, \forall s \in \Sigma^*,$

1. $\tilde{\pi}[\sigma_k, q_j] = 0$ *if* $\delta(q_j, \sigma_k)$ *is undefined; and* $\tilde{\pi}[\epsilon, q_j] = 1$;
2. $\tilde{\pi}[\sigma_k, q_j] \equiv \tilde{\pi}_{jk} \in [0, 1)$; $\Sigma_k \tilde{\pi}_{jk} < 1$;
3. $\tilde{\pi}[\sigma_k, q_j] = \tilde{\pi}[\sigma_k, q_j]\tilde{\pi}[s, \delta(q_j, \sigma k)]$.

The event cost $\tilde{\Pi}$-matrix is given as:

$$\tilde{\Pi} = \begin{bmatrix} \tilde{\pi}_{11} & \tilde{\pi}_{12} & \cdots & \tilde{\pi}_{1n} \\ \tilde{\pi}_{21} & \tilde{\pi}_{22} & \cdots & \tilde{\pi}_{2n} \\ \vdots & & \ddots & \vdots \\ \tilde{\pi}_{n1} & \tilde{\pi}_{n2} & \cdots & \tilde{\pi}_{nn} \end{bmatrix}$$

The condition, $\Sigma_k \tilde{\pi}_{jk} < 1$, in item#2 in Equation (9.3.4), provides a sufficient condition for the existence of the real signed measure [27]; its physical interpretation is given in Chapter 1.

The measure of a sublanguage of the plant language $L(G_i)$ is obtained in terms of the signed characteristic function χ and the event cost matrix $\tilde{\Pi}$, following Chapter 1.

Definition 9.3.5 *Given a DFSA $G_i = (Q, \Sigma, \delta, q_i, Q_m)$, the cost ν of a sublanguage $K \subseteq L(G_i)$ is defined as the sum of the event cost $\tilde{\pi}$ of individual strings belonging to K; that is, $\nu(K) = \Sigma_{s \in k}\tilde{\pi}[s, q_i]$.*

Definition 9.3.6 *The signed real measure μ of a singleton string set $\{s\} \subset L(q_i, q_j) \subseteq L(G_i) \in 2^{\Sigma^*}$ is defined as:*

$$\mu(\{s\}) \equiv x(q_j)\tilde{\pi}(s, q_i)\forall s \in L(q_i, q_j).$$

The signed real measure of the language $L(q_i, q_j)$ is defined as:

$$\mu(L(q_i, q_j)) \equiv \Sigma_{s \in L(q_i, q_j)}\mu(s)$$

and the signed real measure of a DFSA G_i, initialized at state q_i, is defined as:

$$\mu_i \equiv \mu(L(G_i)) = \Sigma_j \mu(L(q_i, q_j)).$$

Definition 9.3.7 *The state transition cost, $\pi : Q \times Q \to [0, 1)$, of a DFSA is defined as follows:*

$$\forall q_i, q_j \in Q, \pi_{ij} = \begin{cases} \Sigma_{\sigma \in \Sigma} \tilde{\pi}[\sigma, q_i], & \text{if } \delta(q_i, \sigma) = q_j \\ 0 & \text{if } \{\delta(q_i, \sigma) = q_j\} = \emptyset. \end{cases} \tag{9.1}$$

Consequently, the $n \times n$ state transition cost Π-matrix is defined as:

$$\Pi = \begin{bmatrix} \pi_{11} & \pi_{12} & \cdots & \pi_{1n} \\ \pi_{21} & \pi_{22} & \cdots & \pi_{2n} \\ \vdots & \vdots & \ddots & \vdots \\ \pi_{n1} & \pi_{n2} & \cdots & \pi_{nn} \end{bmatrix}$$

Wang and Ray [29] have shown that the measure $\mu_i = \mu(L(G_i))$ of the language $L(G_i)$, with the initial state q_i, can be expressed as:

$$\mu_i = \Sigma_j \pi_{ij} \mu_j + \chi_j \text{ where } \chi_i \equiv \chi(q_i)$$

Equivalently, in vector notation:

$$\mu = \Pi\mu + X$$

where $\mu \equiv [\mu_1 \ \mu_2 \ \cdots \ \mu_n]^T$ is the measure vector; and $X = [\chi_1 \ \chi_2 \ \cdots \ \chi_n]^T$ is the the characteristic vector. The solution for $\mu = \Pi\mu + X$ is:

$$\mu = (1 - \Pi)^{-1} X$$

where that $I - \Pi$ is invertible because of the inequality, $\Sigma_k \tilde{\pi}_{jk} < 1$. (See item#2 in Equation (9.3.4)).

The signed real measure given in Definition 9.3.6 for a DFSA is based on the assignment of the characteristic vector and the event cost matrix . As stated earlier, the characteristic function is chosen by the designer based on his/her perception of the states' impact on system performance. On the other hand, the event cost is an intrinsic property of the plant. The event cost $\tilde{\pi}_{jk}$ is conceptually similar to the state transition probability as in Markov Chains, except for the fact that it is not allowed to satisfy the equality condition $\sum_k \tilde{\pi}_{jk} = 1$. (Note that $\Sigma_k \tilde{\pi}_{jk} < 1$ is a requirement for convergence of the language measure.) The rationale for this strict inequality is explained in [27] and also in Chapter 1.

With this interpretation of event cost, $\tilde{\pi}[\sigma, q_i]$ (see Definition 9.3.4) denotes the probability of occurrence of the event string $s \in L(G_i)$ that starts at q_i and terminates at $\delta^*(s, q_i)$. Hence, the cost $\nu(L(q, q_i))$ (see Definition 9.3.5), which is a non-negative real number, is directly related (but not necessarily equal) to the total probability that state q_i would be reached as the plant operates. In the setting of language measure, a supervisor's performance is superior if the supervised plant is more likely to terminate a good marked state and/or less likely to terminate on a bad marked state.

9.4 A Discrete Event Modelling of Malicious Executables

This section describes how the dynamical behavior of malicious executables is modelled as a discrete event system and represent it as a DFSA. Then, the Supervisory Control Theory (SCT) is applied to control the DFSA model such that the detected virus is prevented from spreading into the computer system. The results of the simulation experiments show that the supervised system, on the average, is capable of correctly detecting about 88.75% of malicious executables.

9.4.1 Analysis of Malicious Executables

Though a formal definition of computer virus is still under discussion [25], a commonly accepted understanding of computer virus is that it is a segment of malicious code that can copy itself to other programs when executed. Generally, there are three categories of viruses [17][23][24] as listed below:

- *File viruses* that infect *.exe and *.com files
- *Macro viruses* that infect data files
- *Boot viruses* that infect boot sectors of hard disk or floppy disk.

This chapter focuses on *file virus* which is also called as the malicious executable in this chapter. Generally, a file virus contains a *replicator* that controls the spread of the virus, and a *trigger* that activates the virus so that it can finish its malicious functions [17][23][24]. These malicious functions of the virus may vary and their activation conditions may be different, but their replication and infection strategies are similar. As an extension of the general virus behavior observed in [17] [24], a general outline of replication and infection of a file virus is presented as follows.

```
Procedure: Replication
while not infected enough and files available
{
        find a file;
        if file not infected
        {
                Infection(filename);
                infected file number++;
        }
        else do nothing;
}

Procedure: Infection(file)
{
        Get the file attribute;
        Change the file attribute;
        Save the file date/time stamps;
        Copy virus code to the file;
        Restore the file attribute, date/time stamps;
}
```

This chapter has used an MS-DOS trace tool [26] to trace 25 non-infected programs, and 30 malicious programs created by two virus generation tools:

- BW (http://vx.netlux.org/dat/tb00.shtml).
- G2 (http://vx.netlux.org/dat/tg00.shtml).

The traced system calls verify the general outline given in the procedures *Replication* and *Infection*. A sensitive-sequence is viewed as a sequence of system calls common to most virus programs. An analysis of 30 file virus programs and 25 non-infected programs shows the sensitive-sequence for file virus is: *findfirst, getmod, chmod, open, get_time, read, [write, lseek], set_time, chmod, close*. Note that occurrence of *write* depends on whether the file is infected or not; it may not appear if the file is already infected. The order of the system calls may be slightly different for each virus. For example, *chmod* may happen before or after *open*. Since file viruses spread when the virus code is executing, all system calls of the program are monitored in real time. If the system calls generated by the program follows the sensitive-sequence, it can be predicted with high confidence that the program is either a virus or is infected by a virus.

9.4.2 Synthesis of Plant Model for File Virus

When a process is executing, it interacts with the Operating System (OS). The current system calls, generated by the OS, execute one at a time step to finish the task of the running process. The system calls are modelled as discrete events because they occur asynchronously at discrete instants of time. An executable process is then considered to be a discrete event system that is represented as a DFSA based on the supervisory control theory presented in Section 9.3.1.

Upon creation of a DFSA model of an executable, the next task is to derive the regular language $L(G)$ that is accepted by the model.

$$L(G) = \{s \in \Sigma^* | \delta(q_0, s) \in Q\}$$

and the marked language of G represented as $L_m(G)$, where

$$L_m(G) = \{s \in \Sigma^* | \delta(q_0, s) \in Q\} \subseteq L(G)$$

Figure 9.1 shows the state transition model of a process. The events that trigger the state transition are basically system calls as listed in Table 9.1. As seen in Figure 9.1, the system calls are mapped as events into the alphabet Σ and some of the events may contain more than one system call. In the Table 9.2, where the states are listed, the state 3 is the good marked state where the system wants to terminate on. States 5', 6', 6", 7', 8' and 9' are the bad marked states that the system wants to avoid. State 1 is the idle state, i.e., the state when no process is running. Once the process is started, it transits to state 2. If a transition occurs from state 2 to state, and is followed by state 5',

6' or 6" and 7', then there is a probability that the process contains malicious function; otherwise, it will come to state 3 and finally end at state 1 by event σ_{11}. If it transits to state 8', there is a high probability that the running process contains a malicious code. If it transits to state 9', it is believed that the process contains malicious function. If a user terminates the process once he/she receives the warning message about a (possible) malicious executable, a transition takes place to the state 10 that is a gracefully degraded state because further possible damage is avoided. As such, state 10 is treated as a good marked state with a small positive characteristic value.

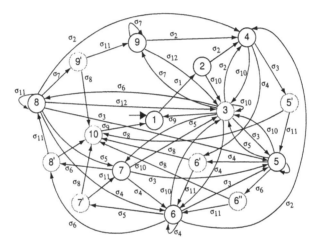

Fig. 9.1. The plant Model (G) of the unsupervised process.

9.4.3 Design and synthesis of Controllers

A supervisor has the same event alphabet as a plant model [22]. Depending on the control policy, the supervisor controls the plant by disabling certain controllable events . The control policy, used here, is to disable the event σ_{11} when it triggers a transition from a bad marked state $q_i \in Q_m^-$ and enables the event σ_8 as the process is terminated upon detecting a virus. At the bad marked states, the supervisory actions are determined by control specifications. The supervisor disables the state transition triggered by some of the controllable events and enables some others to avoid further damage. Consequently, the process arrives at the good marked state 10 and finally ends at state 1. Five controllers were designed corresponding to five different control specifications (see Table 9.4) in Section 9.5.1.

Table 9.1. List of events in the plant model

Event No	Event Name	Type	Description
σ_1	start	controllable	Start executing a process
σ_2	findfirst/findnext getmod chmod open get-time or findfirst/findnext get-data open	uncontrollable	Find the target file
σ_3	read	uncontrollable	Read the file
σ_4	lseek write or open write or lseek or write	uncontrollable	Specify the offset of a file from a certain position; if target not infected infect it
σ_5	set-time or get-time	uncontrollable	Set or get the time stamp
σ_6	close chmod or chmod close or chmod or close	uncontrollable	Close the file; change file mode if necessary
σ_7	chdir	uncontrollable	Change directory
σ_8	terminate	controllable	Terminate running process
σ_9	process-end signal	uncontrollable	A finish executing signal
σ_{10}	common system call	uncontrollable	All the other system calls except the ones that are assigned an event number
σ_{11}	continue execution	controllable	Continue the process without interference

9.5 Simulation and Performance Analysis

This section presents the simulation details, implements the supervisory control system, and evaluates the performance of the unsupervised and supervised plant automata. The performance of different controllers is compared and their relative merits and demerits are assessed. Figure 9.2 presents the structure of the simulation platform, on which the performance evaluation is carried out.

As shown in the lower part of Figure 9.2, the supervisory control system consists of three main modules: (i) *Sensor*; (ii) *Event Generator*; and (iii) *Supervisor*. *Sensor* captures the system calls and user actions when a process is running and transmits these system calls to *Event Generator*. As *Event Generator* receives a system call as an input, it maps the system call(s) into an event and feeds the event to the *Supervisor* to trigger a (possible) state tran-

Table 9.2. List of States in the plant model G

State Number	Description
1	Idle and no processing running
2	Start execution
3	Normal running condition
4	Target file found
5'	File read (decision pending)
5	File Read (continued)
6'	File offset is specified (decision pending)
6"	File mode is changed (decision pending)
6	File is ready to be written (continued)
7'	Time stamp is set/get (decision pending)
7	Time stamp is set/get (continued)
8'	File closed (decision pending)
8	File closed (continued)
9'	Directory is changed (Virus is detected) (decision pending)
9	Directory is changed (Virus is detected) (continued)
10	Execution terminated

sition of the DFSA. A data set has been constructed, consisting of 20 benign executables as well as 60 malicious executables. The malicious executables are generated by three tools: VCL, G2, and BW, that are available at the web site:

$$http://vx.netlux.org/dat/vct.shtml$$

A source file of system calls has been generated for each of the executables in its execution phase with the trace tool. These source files are used to simulate the real executables.

When the simulation begins, *Random Picker* picks an executable randomly from the file set. Once an executable is chosen, a transition automatically takes place from state 1 to state 2. Similarly, when the simulation ends, the process terminates at state 1. When an event is generated, *Event Generator* feeds it to the *Supervisor* to trigger the state transition. A *recorder* is inserted between *Event Generator* and *Supervisor* to trace the event sequence generated for the executing process. As a human user receives the status information from *Supervisor*, he/she may terminate the process, or let it execute without interference. The action of the human user is also captured by the *Sensor* and is mapped as an event (it could be σ_1, σ_8 or σ_{11}) by *Event Generator*. The *Supervisor* enables or disables the event transition and force the process to state 10 according to the control specifications. These steps are shown below.

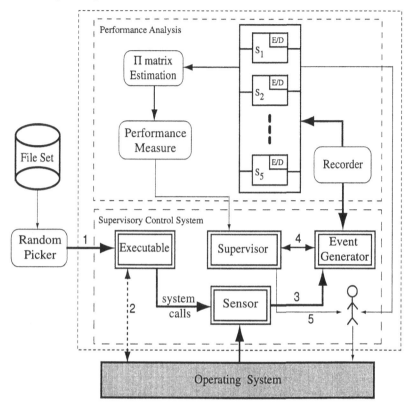

Fig. 9.2. The structure of the simulation process.

Steps:

1. *Random Picker* picks one executable ;
2. Program executing;
3. *Sensor* sends the system calls and user actions to *Event Generator* ;
4. *Event Generator* feeds an event to the *Supervisor* to trigger the state transition;
5. When virus-like behavior is found, *Supervisor* informs user.

Simulation results are summarized in Table 9.3. The frequency of correct detection (i.e.,) is found to be about 88.75%. The frequency of missed detection (i.e., malicious program identified as a normal program) is dependent on the virus creation tool being used. The detection rate for classes of viruses generated by G2 and BW is higher than that for classes of viruses generated by VCL. With the sample programs being equally distributed (i.e., the number of malicious programs generated by each tool and the number of benign programs being the same as in Table 9.3), the frequency of false alarm (i.e., incorrectly identifying presence of a malicious program) is about 10%. The rationale for this high false alarms rate is that the sensitive system calls do not fully cover all categories of viruses. The results are very likely to be improved through more appropriate definition of events.

Table 9.3. Simulation Results

	Benign program	Virus or virus infected programs		
		VCL	BW	G2
No. of programs	20	20	20	20
Detected as virus	2	16	19	18

9.5.1 Performance Analysis

Generally, in the design of a supervisor, one attempts to disable as many bad strings as possible while retaining as many good strings as possible. The main goal of performance analysis, presented in this chapter, is to quantitatively establish the efficacy of supervised processes for detection and prevention of virus spreading, compared to the unsupervised process. If there are several feasible supervisors, the task is to determine which one of the supervisors yields the best performance from the user's perspectives. The procedure of performance analysis and comparison is illustrated on the top part of Figure 9.2 based on the following information.

Recorder generates data files. A data file, called the event file, contains a sequence of events. The Π-Matrix (see Definition 9.3.7) is generated from simulation data using the event files. Performance of the supervisors is then evaluated by combining the Π-Matrix and the χ-vector. The supervisor with the best performance is implemented in the control system.

The supervisory control system has been simulated on the plant automation model with five different controllers as seen in Figure 9.2. Instead of reading the source files, one by one from the file set sequentially, a *Random Picker* is designed to pick the source file of an executable. Each time an executable is picked and fed to the simulation system, the *Recorder* records all events generated until the state 1 is arrived at. In this way, a sequence of events is generated for each executable and this information is stored as an event file. Ten thousand event files were generated for the unsupervised plant automaton and each of the supervised plant automata. The event files have 30% virus data and 70% normal data. This ratio represents the virus encounter probability in the real world [12]. These event files are used to estimate the state transition matrix of the unsupervised and the supervised plant automata.

The language measure parameters (i.e., one χ-vector and six Π-matrices) have been generated for the unsupervised and five supervised plant automata under different supervisors (see Figure 9.2). Numerical values of these language parameters are listed in Appendix A along with the language measure vector μ that is computed by the formula: $\mu = (1 - \Pi)^{-1} X$. The Π-matrix has been identified from the data generated from simulation experiments following the algorithm presented in Chapter 1. Depending on the disabled (controllable) events, the five Π-matrices of the supervised plant automata are determined following the procedure presented in Chapter 2. The weights (i.e., elements of the χ-vector) are selected by the user based the relative importance of the marked states as described below.

Weights are assigned to the marked states in the DFSA based on the role of these states on the plant performance. State 3 is the normal running state that a normal process stays; hence, a positive value is assigned to this state. However, malicious processes may also arrive at this state after the malicious function is finished. Based on the two considerations, a weight of 0.3 is assigned to state 3. State 10 is a pseudo-state where a suspicious process is terminated by the user. Once a process comes to state 10, no further damage is possible; a weight of 0.9 is assigned to state. States 5', 6', 6", 7' 8', 9' are the bad marked state 10; weights of -0.1, -.3, -.3, -0.5, -0.9, -1.0 are assigned to these states, respectively. Notice that the values are decreasing because if a process arrives at these states step by step, the probability that it is a virus becomes higher. In other words, the higher the absolute value of a bad marked state, the worse the situation is. The χ vector thus defined is: $[0\ 0\ 0.3\ 0\ 0\ -0.1\ 0\ -.3\ -.3\ 0\ -0.5\ 0\ -0.9\ 0\ -1.0\ 0.9]^T$ Using the formula $\mu = [I - \Pi]^{-1} \cdot \chi$, it is straight forward to generate the performance vectors of the unsupervised plant automaton and the supervised plant automata (see Appendix A for numerical values of the the language measure vector). Each element μ_i of the μ vector is the performance of the DFSA $G_i \equiv (Q, \Sigma, \delta, q_i, Q_m)$ with initial state q_i. In our model, q_1 is the initial state, so μ_1, the first element of the μ vector, is the performance of the model being evaluated.

The evaluation result is shown in Table 9.4.

Table 9.4. Performance Evaluation Results

	Control Specification (Disabled Event σ_{11})	Performance
Plant Model	none	6.4500
Controller1	State 5' to 5	8.4657
Controller2	State 6' to 6	10.9190
Controller3	State 7' to 7	8.4088
Controller4	State 8' to 8	8.9946
Controller5	State 9' to 9	6.6337

9.5.2 Comparison of Supervisor Performance

Table 9.4 shows that all five supervised plant automata have higher performance than the unsupervised plant automaton. This shows that supervisory control is useful for detection and prevention of virus spreading. A brief discussion on how the five controllers function is presented below.

Controller1 disables state transitions from state 5' to 5; thus it prevents the process from going to the bad states, 6', 6", 7', 8', 9'; but it also hinders the process from going to the good states. Controller2 disables state transition from state 6" to state 6; thus it prevents the process from going to the bad states 7', 8', and 9'. Controller3 disables the state transition from state 7' to state 7; thus it prevents the state from going to the bad state 8' and 9'. Controller4 disables the state transition from state 8' to state 8; the only bad state that is avoided is state 9'. Hence, the performance of Controller4 is relatively low. Controller5 disables the state transition from state 9' to state 9, by which it avoids very few bad states. So, the performance of Controller5 is almost the same as that of the unsupervised plant automaton.

The detection frequency depends on the events and sequence of a priori defined system calls, which make the state transitions of lead to the bad marked states. The following paragraph presents performance analysis.

A false alarm, i.e., a normal program being identified as a malicious one, may have detrimental effects, such as user-initiated termination, or ignoring future warning messages on arrival of truly malicious programs. The supervisor is designed to steer the process to the good marked state 10, even if the user may choose to discard the warning message. As explained in Section 9.3.2, a supervisor's performance is superior if the controlled plant is more likely to reach one of the good marked state and/or less likely to terminate on one of the bad marked states. In this case, the controller is designed to block malicious programs and to allow uninterrupted operation of normal programs. In other words, the DFSA be brought to state 3 or state 10. However, in blocking a state transition to a bad marked state, a good marked state may also be blocked. For example, if Controller#1 disables a state transition to any one of the bad states, 6', 6", 7', 8', 9', the states 6, 7, and 8, which lead to the good marked state 3, may become unreachable. So, there is a trade-off between avoidance of bad marked states and access to good marked states. The language measure theory, presented in Chapter 1, provides a quantitative way to design discrete-event controllers. Table 9.4 shows that Controller#2 has the highest performance among the five supervisory controllers.

9.6 Conclusions

This chapter presents modelling and simulation of a process execution as a discrete event system that can be represented as a Deterministic Finite State Automaton (DFSA). The application of the supervisory control theory (SCT) is shown on a simulation test bed that spreading of malicious executables can be mitigated from the perspectives of both correct detection and false alarms. The test case are based on the file virus as an example. Five supervisory controllers have been designed based on five different control specifications. The language measure theory, presented in Chapter 1 is applied to compare the performance of the five supervised process automata. The language measure parameters are evaluated from the simulation test data.

Although The reported work demonstrate the efficacy of discrete-event supervisory (DES) control to detect and prevent spreading of malicious executables, it is preliminary in nature. Further work is necessary before any firm conclusions can be drawn regarding applicability of the proposed method for commercial applications. From this perspective, future research is recommended in the following areas: (i) analytical work for extension to higher level languages in the Chomsky hierarchy (see Chapter 4; (ii) capturing the system calls real-time; and (iii) implementation of the supervisory control system in real word to detect the control the malicious executables on-line.

Appendix A
Language Measure Parameters and Performance

This appendix presents the parameters (i.e., the χ-vector and Π-matrices) and the computed language measure μ for the unsupervised process automaton and five supervised process automata under different specifications. The χ-vector that is invariant for the unsupervised plant automaton and all five supervised plant automata is selected as:

$$\chi = [0.0\ \ 0.0\ \ 0.3\ \ 0.0\ \ 0.0\ \ \text{-}0.1\ \ 0.0\ \ \text{-}0.3\ \ \text{-}0.3\ \ 0.0\ \ \text{-}0.5\ \ 0.0\ \ \text{-}0.9\ \ 0.0\ \ \text{-}1.0\ \ 0.9]^T.$$

The language measure vector is computed as:

$$\mu = (1 - \Pi)^{-1} X$$

Since state 1 is the initial state, only the the first element, μ_1, of the μ-vector is relevant. The results are summarized in Table 9.4. Following are the Π-matrices and μ vectors:

Π-Matrix for the unsupervised process automaton

$$
\begin{bmatrix}
0 & .980 & 0 & 0 & 0 & 0 & 0 & 0 & 0 & 0 & 0 & 0 & 0 & 0 & 0 & 0 \\
0 & 0 & .936 & .044 & 0 & 0 & 0 & 0 & 0 & 0 & 0 & 0 & 0 & 0 & 0 & 0 \\
.039 & 0 & .799 & .009 & .027 & 0 & .042 & 0 & 0 & .018 & 0 & .009 & 0 & .036 & 0 & 0 \\
0 & 0 & 0 & 0 & 0 & .931 & 0 & .049 & 0 & 0 & 0 & 0 & 0 & 0 & 0 & 0 \\
0 & 0 & .138 & 0 & .244 & 0 & 0 & .534 & .064 & 0 & 0 & 0 & 0 & 0 & 0 & 0 \\
0 & 0 & 0 & 0 & .980 & 0 & 0 & 0 & 0 & 0 & .156 & 0 & 0 & 0 & 0 & 0 \\
0 & 0 & .260 & .007 & .214 & 0 & .032 & 0 & 0 & 0 & 0 & 0 & .019 & 0 & 0 & 0 \\
0 & 0 & 0 & 0 & 0 & 0 & .980 & 0 & 0 & 0 & 0 & 0 & 0 & 0 & 0 & 0 \\
0 & 0 & 0 & 0 & 0 & 0 & .980 & 0 & 0 & 0 & 0 & 0 & 0 & 0 & 0 & 0 \\
0 & 0 & .15 & 0 & .00050 & 0 & .074 & 0 & 0 & 0 & 0 & 0 & .752 & 0 & 0 & 0 \\
0 & 0 & 0 & 0 & 0 & 0 & 0 & 0 & 0 & .980 & 0 & 0 & 0 & 0 & 0 & 0 \\
0 & 0 & .608 & .186 & .049 & 0 & .046 & 0 & 0 & .0004 & 0 & .061 & 0 & 0 & .031 & 0 \\
0 & 0 & 0 & 0 & 0 & 0 & 0 & 0 & 0 & 0 & 0 & .980 & 0 & 0 & 0 & 0 \\
0 & 0 & .879 & .055 & 0 & 0 & 0 & 0 & 0 & 0 & 0 & 0 & 0 & .047 & 0 & 0 \\
0 & 0 & 0 & 0 & 0 & 0 & 0 & 0 & 0 & 0 & 0 & 0 & 0 & .980 & 0 & 0 \\
0 & 0 & 0 & 0 & 0 & 0 & 0 & 0 & 0 & 0 & 0 & 0 & 0 & 0 & 0 & 0
\end{bmatrix}
$$

μ-vector for the unsupervised process automaton

$$[6.45\ 6.58\ 6.79\ 5.02\ 5.33\ 5.12\ 5.58\ 5.17\ 5.17\ 0.29\ 4.689\ 6.12\ 5.10\ 6.55\ 5.42\ 1.00]^T$$

Π-Matrix for the supervised process automaton under Controller1

```
⎡ 0    .980  0     0     0     0    0     0    0    0     0  0     0    0    0  0    ⎤
⎢ 0     0   .935  .045   0     0    0     0    0    0     0  0     0    0    0  0    ⎥
⎢.035   0   .808  .006  .032   0   .043   0    0   .019   0 .010   0   .028  0  0    ⎥
⎢ 0     0    0     0     0    .890  0    .090  0    0     0  0     0    0    0  0    ⎥
⎢ 0     0   .182   0    .268   0    0    .490 .037  0     0  0     0    0    0  0    ⎥
⎢ 0     0    0     0     0     0    0     0    0    0     0  0     0    0    0 .980  ⎥
⎢ 0     0   .324  .276   0    .317  0     0    0   .045   0 .018   0    0    0  0    ⎥
⎢ 0     0    0     0     0     0   .980   0    0    0     0  0     0    0    0  0    ⎥
⎢ 0     0    0     0     0     0   .980   0    0    0     0  0     0    0    0  0    ⎥
⎢ 0     0   .267   0    .002   0    0     0    0    0     0  0    .710  0    0  0    ⎥
⎢ 0     0    0     0     0     0    0     0    0   .980   0  0     0    0    0  0    ⎥
⎢ 0     0   .91   .001  .095   0   .049   0    0   .002   0 .046   0    0  .002 0    ⎥
⎢ 0     0    0     0     0     0    0     0    0    0     0 .980   0    0    0  0    ⎥
⎢ 0     0   .91   .041   0     0    0     0    0    0     0  0     0   .029  0  0    ⎥
⎢ 0     0    0     0     0     0    0     0    0    0     0  0     0   .980  0  0    ⎥
⎣.980   0    0     0     0     0    0     0    0    0     0  0     0    0    0  0    ⎦
```

μ-vector for the supervised process automaton under Controller1

$$[8.47\ 8.64\ 8.82\ 8.69\ 7.58\ 9.01\ 7.9\ 7.44\ 7.44\ 7.62\ 6.97\ 8.46\ 7.39\ 8.63\ 7.46\ 9.30]^T$$

Π-Matrix for the supervised process automaton under Controller2

```
⎡ 0    .980  0     0     0     0    0     0    0    0     0  0     0    0    0  0    ⎤
⎢ 0     0   .937  .042   0     0    0     0    0    0     0  0     0    0    0  0    ⎥
⎢.030   0   .847  .008  .005   0   .037   0    0   .007   0 .011   0   .034  0  0    ⎥
⎢ 0     0    0     0     0    .918  0    .062  0    0     0  0     0    0    0  0    ⎥
⎢ 0     0   .088   0    .223   0    0    .452 .217  0     0  0     0    0    0  0    ⎥
⎢ 0     0    0     0    .980   0    0     0    0    0     0  0     0    0    0  0    ⎥
⎢ 0     0   .366  .018  .163   0   .398   0    0    0     0 .036   0    0    0  0    ⎥
⎢ 0     0    0     0     0     0    0     0    0    0     0  0     0    0    0 .980  ⎥
⎢ 0     0    0     0     0    .980  0     0    0    0     0  0     0    0    0  0    ⎥
⎢ 0     0   .831   0    .011   0    0     0    0    0     0  0    .138  0    0  0    ⎥
⎢ 0     0    0     0     0     0    0     0    0   .980   0  0     0    0    0  0    ⎥
⎢ 0     0   .839   0     0     0    0     0    0   .008   0 .133   0    0    0  0    ⎥
⎢ 0     0    0     0     0     0    0     0    0    0     0 .980   0    0    0  0    ⎥
⎢ 0     0   .935  .045   0     0    0     0    0    0     0  0     0    0    0  0    ⎥
⎢ 0     0    0     0     0     0    0     0    0    0     0  0     0    0    0  0    ⎥
⎣.980   0    0     0     0     0    0     0    0    0     0  0     0    0    0  0    ⎦
```

μ-vector for the supervised process automaton under Controller2

$$[10.92\ 11.14\ 11.42\ 10.17\ 10.64\ 0.33\ 10.73\ 11.17\ 10.21\ 10.99\ 10.27\ 11.16\ 10.03\ 11.14$$
$$-1.00\ 11.70]^T$$

Π-Matrix for the supervised process automaton under Controller3

$$\begin{bmatrix}
0 & .980 & 0 & 0 & 0 & 0 & 0 & 0 & 0 & 0 & 0 & 0 & 0 & 0 & 0 & 0 \\
0 & 0 & .936 & .044 & 0 & 0 & 0 & 0 & 0 & 0 & 0 & 0 & 0 & 0 & 0 & 0 \\
.031 & 0 & .810 & .008 & .034 & 0 & .038 & 0 & 0 & .021 & 0 & .008 & 0 & .031 & 0 & 0 \\
0 & 0 & 0 & 0 & .921 & 0 & .059 & 0 & 0 & 0 & 0 & 0 & 0 & 0 & 0 & 0 \\
0 & 0 & .189 & 0 & .221 & 0 & 0 & .059 & .061 & 0 & 0 & 0 & 0 & 0 & 0 & 0 \\
0 & 0 & 0 & 0 & .980 & 0 & 0 & 0 & 0 & 0 & 0 & 0 & 0 & 0 & 0 & 0 \\
0 & 0 & .298 & .009 & .251 & 0 & .322 & 0 & 0 & 0 & .078 & 0 & .022 & 0 & 0 & 0 \\
0 & 0 & 0 & 0 & 0 & 0 & .980 & 0 & 0 & 0 & 0 & 0 & 0 & 0 & 0 & 0 \\
0 & 0 & 0 & 0 & 0 & 0 & .980 & 0 & 0 & 0 & 0 & 0 & 0 & 0 & 0 & 0 \\
0 & 0 & .312 & 0 & .002 & 0 & 0 & 0 & 0 & 0 & 0 & 0 & .666 & 0 & 0 & 0 \\
0 & 0 & 0 & 0 & 0 & 0 & 0 & 0 & 0 & 0 & 0 & 0 & 0 & 0 & 0 & .980 \\
0 & 0 & .837 & 0 & 0 & 0 & .087 & 0 & 0 & .002 & 0 & .055 & 0 & 0 & 0 & 0 \\
0 & 0 & 0 & 0 & 0 & 0 & 0 & 0 & 0 & 0 & 0 & .980 & 0 & 0 & 0 & 0 \\
0 & 0 & .942 & .038 & 0 & 0 & 0 & 0 & 0 & 0 & 0 & 0 & 0 & 0 & 0 & 0 \\
0 & 0 & 0 & 0 & 0 & 0 & 0 & 0 & 0 & 0 & 0 & 0 & 0 & 0 & 0 & 0 \\
.980 & 0 & 0 & 0 & 0 & 0 & 0 & 0 & 0 & 0 & 0 & 0 & 0 & 0 & 0 & 0
\end{bmatrix}$$

μ-vector for the supervised process automaton under Controller3

$$[8.41 \quad 8.58 \quad 8.82 \quad 7.30 \quad 7.69 \quad 7.44 \quad 8.05 \quad 7.59 \quad 7.59 \quad 7.76 \quad 8.56 \quad 8.56 \quad 7.49 \quad 8.59 \quad -1.00 \quad 9.24]^T$$

Π-Matrix for the supervised process automaton under Controller4

$$\begin{bmatrix}
0 & .980 & 0 & 0 & 0 & 0 & 0 & 0 & 0 & 0 & 0 & 0 & 0 & 0 & 0 & 0 \\
0 & 0 & .941 & .039 & 0 & 0 & 0 & 0 & 0 & 0 & 0 & 0 & 0 & 0 & 0 & 0 \\
.030 & 0 & .846 & .008 & .008 & 0 & .039 & 0 & 0 & .009 & 0 & .010 & 0 & .031 & 0 & 0 \\
0 & 0 & 0 & 0 & 0 & .931 & 0 & .049 & 0 & 0 & 0 & 0 & 0 & 0 & 0 & 0 \\
0 & 0 & .099 & 0 & .180 & 0 & 0 & .583 & .117 & 0 & 0 & 0 & 0 & 0 & 0 & 0 \\
0 & 0 & 0 & 0 & .980 & 0 & 0 & 0 & 0 & 0 & 0 & 0 & 0 & 0 & 0 & 0 \\
0 & 0 & .227 & .011 & .283 & 0 & .351 & 0 & 0 & 0 & .081 & 0 & .027 & 0 & 0 & 0 \\
0 & 0 & 0 & 0 & 0 & 0 & .980 & 0 & 0 & 0 & 0 & 0 & 0 & 0 & 0 & 0 \\
0 & 0 & 0 & 0 & 0 & 0 & .980 & 0 & 0 & 0 & 0 & 0 & 0 & 0 & 0 & 0 \\
0 & 0 & .367 & 0 & .003 & 0 & .018 & 0 & 0 & 0 & 0 & 0 & .592 & 0 & 0 & 0 \\
0 & 0 & 0 & 0 & 0 & 0 & 0 & 0 & 0 & .980 & 0 & 0 & 0 & 0 & 0 & 0 \\
0 & 0 & .837 & 0 & 0 & 0 & 0 & 0 & 0 & .005 & 0 & .138 & 0 & 0 & 0 & .0 \\
0 & 0 & 0 & 0 & 0 & 0 & 0 & 0 & 0 & 0 & 0 & 0 & 0 & 0 & 0 & .980 \\
0 & 0 & .930 & .050 & 0 & 0 & 0 & 0 & 0 & 0 & 0 & 0 & 0 & 0 & 0 & 0 \\
0 & 0 & 0 & 0 & 0 & 0 & 0 & 0 & 0 & 0 & 0 & 0 & 0 & 0 & 0 & 0 \\
.980 & 0 & 0 & 0 & 0 & 0 & 0 & 0 & 0 & 0 & 0 & 0 & 0 & 0 & 0 & 0
\end{bmatrix}$$

μ-vector for the supervised process automaton under Controller4

$$[8.99 \quad 9.18 \quad 9.45 \quad 7.34 \quad 7.74 \quad 7.48 \quad 8.18 \quad 7.71 \quad 7.71 \quad 8.80 \quad 8.12 \quad 9.23 \quad 8.72 \quad 9.15 \quad -1.00 \quad 9.81]^T$$

Π-Matrix for the supervised process automaton under Controller5

$$\begin{bmatrix}
0 & .980 & 0 & 0 & 0 & 0 & 0 & 0 & 0 & 0 & 0 & 0 & 0 & 0 & 0 & 0 \\
0 & 0 & .938 & .042 & 0 & 0 & 0 & 0 & 0 & 0 & 0 & 0 & 0 & 0 & 0 & 0 \\
.036 & 0 & .806 & .008 & .026 & 0 & .039 & .041 & 0 & .017 & 0 & .010 & 0 & .040 & 0 & 0 \\
0 & 0 & 0 & 0 & .940 & 0 & .555 & 0 & 0 & 0 & 0 & 0 & 0 & 0 & 0 & 0 \\
0 & 0 & .134 & 0 & .235 & 0 & 0 & 0 & .056 & 0 & 0 & 0 & 0 & 0 & 0 & 0 \\
0 & 0 & 0 & 0 & .980 & 0 & 0 & 0 & 0 & 0 & 0 & 0 & 0 & 0 & 0 & 0 \\
0 & 0 & .247 & .005 & .226 & 0 & .321 & 0 & 0 & 0 & .160 & 0 & .021 & 0 & 0 & 0 \\
0 & 0 & 0 & 0 & 0 & 0 & .980 & 0 & 0 & 0 & 0 & 0 & 0 & 0 & 0 & 0 \\
0 & 0 & 0 & 0 & 0 & 0 & .980 & 0 & 0 & 0 & 0 & 0 & 0 & 0 & 0 & 0 \\
0 & 0 & .157 & 0 & .001 & 0 & .079 & 0 & 0 & 0 & 0 & 0 & .743 & 0 & 0 & 0 \\
0 & 0 & 0 & 0 & 0 & 0 & 0 & 0 & 0 & .980 & 0 & 0 & 0 & 0 & 0 & 0 \\
0 & 0 & .604 & .198 & .043 & 0 & .047 & 0 & 0 & .001 & 0 & 0 & 0 & .037 & 0 & 0 \\
0 & 0 & 0 & 0 & 0 & 0 & 0 & 0 & 0 & 0 & 0 & .052 & 0 & 0 & 0 & 0 \\
0 & 0 & .886 & .057 & 0 & 0 & 0 & 0 & 0 & 0 & 0 & .980 & 0 & .037 & 0 & 0 \\
0 & 0 & 0 & 0 & 0 & 0 & 0 & 0 & 0 & 0 & 0 & 0 & 0 & 0 & 0 & .980 \\
.980 & 0 & 0 & 0 & 0 & 0 & 0 & 0 & 0 & 0 & 0 & 0 & 0 & 0 & 0 & 0
\end{bmatrix}$$

μ-vector for the supervised process automaton under Controller5

$[6.63\ 6.77\ 6.99\ 5.13\ 5.45\ 5.24\ 5.70\ 5.28\ 5.28\ 5.47\ 4.86\ 6.29\ 5.27\ 6.73\ 6.35\ 7.50]^T$

References

1. S. T. Eckmann, G. Vigna, and R.A. Kemmerer, *Statl: An attack language for state-based intrusion detection*, Proceedings of the ACM workshop on Intrusion Detection (Athens, Greece), November 2000.
2. E. Eskin and W. Lee, *Modeling system calls for intrusion detection with dynamic window sizes*, Proceedings of DISCEX II, June 2001.
3. S. Forrest, S. A. Hofmeyr, and A. Somayaji, *Intrusion detection using system calls*, Journal of computer society **6** (1998), no. 3, 151–180.
4. S. Forrest, S. A. Hofmeyr, A. Somayaji, and T. A. Longstaff, *A sense of self for unix processes*, Proceedings of the 1996 IEEE Symposium on Security and Privacy, 1996, pp. 120–128.
5. S. Forrest, A.S. Perlelson, L. Allen, and R. Cherukuri, *Self-noneself discrimination in a computer*, Proceedings of IEEE Symposium on research in Security and Privacy, 1994.
6. S. Forrest, C. Warrender, and B. Pearlmutter, *Detecting intrusion using system calls: Alternative data models*, Proceedings of the 1999 IEEE Symposium on Security and Privacy, 1999, pp. 133–145.
7. A.K. Ghosh, A. Schwartzbard, and M. Schatz, *Learning program behavior profiles for intrusion detection*, Proceedings of the 1st USENIX Workshop on Intrusion Detection and Networking Monitoring, 1999.
8. K. Goseva-Popstojanova, F. Wang, R. Wang, F. Gong, K. Vaidyanathan, K. Trivedi, and B.Muthusamy, *Characterizing intrusion tolerant systems using a state transition model*, Proceedings of DARPA Information Survivability Conference and Exposition II (DISCEX'01), 2001.
9. K. Ilgun, *Ustat: A real time intrusion detection system for unix*, Proceedings of the 1993 IEEE Symposium on Research in Security and Privacy, 1993.
10. K. Ilgun, R. A. Kemmerer, and P. A. Porras, *State transition analysis: a rule-based intrusion detection approach*, IEEE Transactions on software engineering **21** (1995), no. 3.
11. G. Karsai and A. Ledeczi, *An approach to self adaptive software based on supervisory control*, IWSAS 2001 (Balatonfured, Hungary), 2001.
12. J.O. Kephart and S.R. White, *How prevalent are computer viruses,* High Integrity Computing Laboratory, http://www.research.ibm.com/antivirus/SciPapers/Kephart/DPMA92/dpma92.html.
13. R. Kumar and M. Fabian, *Supervisory control of partial specification arising in protocol conversion*, 35th Allerton Conference on Communication, Control and computing (Urbana-Champaign, Illinois), 1997, pp. 543–552.
14. S. Kumar and E. H. Spafford, *A generic virus scanner in c++*, Proceedings of the 8th Computer Security Applications Conference, IEEE press, 1992.
15. B. LeCharlier and M. Swimmer, *Dynamic detection and classification of computer viruses using general behavior patterns*, Proceedings of Fifth International Virus Bulletin Conference, September 20-22 1995, p. 75.

16. W. Lee and S.J. Stolfo, *Data mining approaches for intrusion detection*, Proceedings of the Seventh USENIX Security Symposium (SECURITY '98) (San Antonio, TX), January 1998.
17. R.B. Levin, *The computer virus handbook*, Osbrne McGraw-Hill, 1990, ISBN 0-07-881047-5.
18. C.C. Michael and A. Ghosh, *Using finite automata to mine execution data for intrusion detection: a preliminary report*, Proc. RAID 2000, 2000, pp. 133–145.
19. N. Nuansri, S. Singh, and T.S. Dillon, *A process state-transition analysis and its application to intrusion detection*, ACSAC 1999, 1999, pp. 378–388.
20. Y. Okazaki, I. Sato, and S. Goto, *A new intrusion detection method based on process profiling*, Proceedings of the 2002 Symposium on Applications and the Internet (SAINT) (Nara City, Nara, Japan), Jan 28 - Feb 01 2002.
21. V. Phoha, X. Xu, A. Ray, and S. Phoha, *Supervisory control automata paradigm to make malicious executables ineffectual*, Proceedings of the 5th IFAC Symposium on Fault Detection, Supervision and Safety (Washington, D.C), June 2003.
22. P.J. Ramadge and W.M. Wonham, *Supervisory control of a class of discrete event processes*, SIAM J. Control and Optimization **25** (1987), no. 1, 206–230.
23. M. G. Schultz, E. Eskin, E. Zadok, and S.J. Stolfo, *Data mining methods for detection of new malicious executables*, Proceedings of IEEE Symposium on Security and Privacy (Oakland, CA), May 2001.
24. A. Solomon and T. Kay, *Dr. solomon's pc anti-virus book*, Newtech, 1994, ISBN 0750616148.
25. E.H. Spafford, *Computer viruses as artificial life*, Journal of Artificial life **1** (1994), no. 3, 249–265.
26. D. Spinellis, *Trace: A tool for logging operating system call transactions*, Operating Systems Review **28** (1994), no. 4, 56–63.
27. A. Surana and A. Ray, *Signed real measure of regular languages*, Demonstratio Mathematica **37** (2004), no. 2, 485–503.
28. C. Wallace, P. Jensen, and N. Soparkar, *Supervisory control of workflow scheduling*, Proceedings of International Workshop on Advanced Transaction Models and Architectures (Goa), August - September 1996.
29. X. Wang and A. Ray, *A language measure for performance evaluation of discrete-event supervisory control systems*, Applied Mathematical Modelling **28** (2004), no. 9, 817–833.

Index